A MANUAL FOR DIVERSIFYING YOUR FARM INCOME

MATT STEPHENS

1503 S.W. 42nd St.
Topeka, KS 66609-1265, USA
Telephone: (785) 274-4300
(800) 678-5779
Fax: (785) 274-4305
www.ogdenpubs.com

Publisher and Editorial Director: Bill Uhler
Director of Events and Business Development: Andrew Perkins
Production Director: Bob Cucciniello
Special Content Group Editor: Jean Teller
Special Content Copy Editor: Karmin Garrison
Book Design and Layout: Laura Perkins

Disclaimer: The content of this book is educational and not intended to be a substitute for advice from licensed legal or tax professionals. Do your research and find a legal and/or tax professional you can work with for your specific needs. The views expressed are the author's and the farmers he interviewed, and neither the author or Ogden Publications shall have liability nor responsibility to any person or entity with respect to any loss, damage, or injury caused or alleged to be caused directly or indirectly by the information contained in this book.

Ogden Publication titles are also available at discount for retail, wholesale, promotional, and bulk purchase. For details, contact Customer Service: (800) 234-3368; email at customerservice@ogdenpubs.com; or by mail at Ogden Publications, Inc., 1503 S.W. 42nd St., Topeka, KS 66609-1265.

ISBN: 978-1-948734-26-4

Printed in the U.S.A.

DEDICATION

This book is dedicated to:

- David Fredrickson for always supporting my dreams and picking up the phone.
- Brian and Helen Robinson for helping me to "keep moving forward."
- All the people holding this book with a farm dream in their hearts.

ACKNOWLEDGMENTS

Thank you to all the farmers I formally interviewed for this book; read more about them and their operations beginning on page 10. And to all the people I've talked with over the years who shared their stories and their passion for helping others experience the rural life.

My appreciation goes to the many friends and family members who suffered with me to bring this book to life; specifically, my wife, Summer, and my writing mentor John.

A few people tried to discourage me from writing this book and I thank them. Their words gave me the fuel to push through and reach the people who now hold this book in their hands.

As a public educator, I applaud all who teach. Especially my first teachers who endured my presence in their classrooms. These brave souls built the foundation for these words. Who would we be without the true teachers in our lives?

No agritourism business would be in operation without the people and families who visit our farms and will visit your farms in the future.

To bring you these pages, I thank all my Fair family with the Mother Earth News Fairs and the publishing team at Ogden Publications.

MY PERSONAL VISION STATEMENT

To help the world by sharing my story, successes and failures alike.

MY BUSINESS VISION STATEMENT

No farm dream should fail.

MISSION STATEMENT FOR THIS BOOK

"All ships come up in a rising tide."

—Valdeane W. Brown, Zengar Institute,
Victoria, BC, Canada, "Neurofeedback: The First Fifty Years,"
edited by James R. Evans, Mary Blair Dellinger, and Harold L. Russell
(2020, Academic Press, an imprint of Elsevier)

I want this community of agritourism to be that tide for family and small farms. I want to start a movement and help 1 million farms succeed by adding or growing with agritourism. This book is to intrigue and inspire you to know you can do it. I want to bring like-minded people together for the common good of all family and small farms.

If you find inspiration in these pages, please contact me directly at matt@stephensfamily.farm.

TABLE OF CONTENTS

"It is only through our collective wisdom that we can survive."

—**Matt Stephens**

INTRODUCTION

I was in the audience the day a farmer was describing a school field trip on his farm. He told of how he reached under a hen in front of the group and pulled out an egg. A young voice rang out with a question, "Who put that there?" The tour continued to the milking barn. As their guide squeezed a handful of milk from one of the milking cows, another question was posed, "What will they think of next?" The question would not have been quite so disturbing except it came from the teacher.

Sitting there that day, I decided to take my farm public and educate the masses.

I believe all farm dreams can succeed and all farms need agritourism. And this is why:

We all have a story to tell. We all can educate people who experience our operations. You can bring people onto your farm and teach them about your passion, whether it's wool, tomatoes, or the natural history of your area. Even if you don't think yourself capable, we will discuss ways to painlessly do most aspects of agritourism.

Some concepts of this book will overlap into different areas of agritourism. As you read through these pages, I want you to think of your ideal farmstead setup. What does it look like? What are the smells that waft on the breeze and the sounds that drift from the pastures? What does it feel like to own this first-rate property?

What is Agritourism?

Agritourism has many definitions. It depends on where you are getting your information and who you're talking to. When you start researching the term "agritourism," it seems there is a different definition for every individual operation, and this is great because it means agritourism is completely customizable to your farm. The word "agritourism" combines "agriculture" and "tourism." Dictionaries define it by breaking down the parts of the word. Online sources can't even agree on how to spell the word: "agritourism" or "agrotourism."

In defining agritourism, you also must ask, "To whom are you speaking?" Is it an insurance agent, lawyer, or your accountant? All could have different definitions and opinions on the subject. You may have to change agents, lawyers, or tax professionals because they either don't handle this type of ag-business or can't. These individuals are key to any successful endeavor. You have to follow their guidelines to operate your business. We'll talk more about the need for professionals in later chapters when we get down to the business of agritourism.

So we need to set our working definition for this book. Number one is spelling; yes, I am a recovering public school teacher. Agritourism = Ag + grit + tourism. *Ag* is for agriculture. *Grit* is for the tenacity it takes to be successful. *Tourism* is for all the people who want to see what you have to offer.

After traveling the country and touring many different types and styles of operations, my definition is simple: For me, "agritourism" means "inviting people to experience your farm dream." Some of the operations I've seen charge for people to come onto the farm. Other operations make enough money growing corn or other crops or as a working cattle ranch that they don't charge anything; some take donations. As a hybrid, a few allow special groups to come out for free while charging the masses. By this definition, you could have a 1-acre pick-your-own, a half-acre market garden, a 50-acre pumpkin patch, a hundred acres of Christmas trees, or the ability to invite people into your city home to teach them how to spin fibers from locally sourced farms. For me, all of these are "agritourism" because you are sharing your agri-dream with others. I have a broad definition of "agritourism" because I've been practicing agritourism for more than a decade—speaking on the topic, doing farm tours, and interviewing many operators. That being said, most of us dirt-turners do not make enough money on our farms to break dirt for free.

The focus of this book will be on the income and profit-generating section of agritourism. Money is not a bad thing; it is a tool like a shovel, and income makes a farm sustainable.

What isn't Agribusiness?

Whether it's keeping a flower bed weed-free or managing a thousand-acre spread, nothing in agriculture is easy. Each size operation has its own unique ups and downs with extra labor needed the larger you get. Agritourism is no different; it's not always easy. If you enjoy working, then you will enjoy the work, but you will not enjoy all aspects of agritourism duties. Part of the purpose of this book is to set realistic expectations. If you enjoy a 9-to-5 work schedule, then agritourism may not be for you. Even if you have only an annual event, there will be year-round tasks that need to happen for that event to be a success. And just because your website says a certain time and date, the public will show up any time they feel like it.

Who is this book for?
Anyone with a farm dream.

What is agritourism?
People and a farm experience coming together.

Where is agritourism?
Anywhere people and agriculture can interact.

When is the time for agritourism?
Year-round—24/7/365.

Why agritourism?
To educate and foster sustainability in a fun way.

How can you be successful?
This book will guide you.

"A broke farmer helps no one but the big food industries."

—Matt Stephens

FARMER & FARM PROFILES

Getting to know your people.

BONNIE CHAPA

Laughing Llama Farm – Troy, TX

Bonnie began her farm story while working in the real estate industry. She and her husband, Frank, moved back to her hometown of Troy, Texas, about 30 minutes away from Waco. They had to wait an entire decade before the property they wanted became available. They approached the owner and were able to make a deal; Bonnie says they technically traded houses. The farm owners were looking to downsize, and Bonnie and Frank were looking to upsize into a dream property.

When they acquired the property, it was not being used as a farm and things were rundown. They had to have a place to live, so they started remodeling the house and making plans for various stages of development.

They settled on raising sheep. The family specifically picked Dorper hair sheep that could add income. In their research, they also discovered the many predators that created a whole new list of challenges. Bonnie was always fascinated by llamas and alpacas and, in doing her homework on predation, found that if they were going to raise sheep, they needed a guard animal for the herd.

Among the options were livestock guardian dogs, donkeys, and emus, but the one species that kept popping up was the llama. About that time, Bonnie met a neighbor who was downsizing her llama herd; they struck a deal, and, the next thing she knew, she was the owner of a five-llama herd.

During this time, their two sons—at the time, Landon was 11 and Logan was 14—made a deal with Bonnie's dad. Grandpa fronted the seed money for a herd of Dorper sheep and all the equipment. The boys would manage their own herd of sheep, and Mom and Dad would manage the llamas. The five-year agreement with Grandpa had to be paid back. In less than two years, the boys honored their agreement; they have played a key role in the overall success of the farm.

When we talked, Bonnie spoke about people's dreams of making their homestead into a business and how most people work a 'real job' with a daily 9-to-5 grind while trying to make the homestead work for them.

In her case, she works remotely from home with a job flexible enough to allow her to work the homestead as well as their agritourism business. This is when we started talking about Bonnie's philosophy on agritourism and breaking it into different branches. Some of the areas included outdoor recreation like fishing, nature walks, and horseback riding. Then we delved into educational events like cooking classes, wine tastings, or how-to workshops.

With the success of the livestock, it was time for a new project. Bonnie and Frank started talking about short-term rental options. Their studies included the cost of different types of buildings and that's when she got the idea of renovating an old grain bin into the Silo House.

For another year, Bonnie did her homework on converting a grain bin or silo, even though they didn't have a standing silo on the property. They decided to reclaim two bins they found 50 miles away in the middle of a cornfield. They had to wait until the corn was combined, but then weather set in. This particular August brought horrendous rains and the rows were running like rivers. In the meantime, the engineering and slab work were being done back on the homestead for one of the bins to become rental property.

The process started in mid-August 2018 and, by mid-January 2019, the silo had its first guests. When the initial visitors arrived, the llamas were in the pastures, people could go fishing in the tank, and the whole dream of The Silo House at Laughing Llama Farm was a reality.

FRAN TACY

Franny's Farm–Asheville, NC

Fran's farm story began when she was born into an agriculture and business family; her dad was a cattle rancher and her mom was in the corporate world. Both were well educated with degrees in business and theology. Fran grew up riding horses and working on her dad's farm. For her mother, she would plant a garden on Mother's Day because that was when it was safe to plant in their region.

Fran went to forestry school, which, at the time, was similar to a sustainable agriculture curriculum, though it was before "sustainable ag" was popular. She lived and worked in the woods for the Forestry Service every summer while holding onto her childhood dream of having her own farm.

She knew she wanted a family and to homeschool her children with a garden to feed the entire household. With that clear picture in mind, she decided to leave the Forestry Service. She went back to college, earned a master's degree in education, and started teaching school. She used a garden to help teach real-life skills to her special education students, and she still dreamed of replicating her childhood farm.

Ultimately, Fran purchased the land for her dream and started to build every structure of the farm. Along the way, she met several mentors and helped to change city ordinances regarding backyard chickens in Asheville, North Carolina.

Fran worked on genetics for heritage poultry for nearly a decade and taught classes on what she learned. In 2017, while she was teaching poultry and business workshops, she started organizing to grow hemp. Fran became the first female industrial hemp farmer in North Carolina.

Fran's agritourism story started after the farm had been operating for four years, and she had been on-site full-time for a year. That's when they built the short-term rental cabins. Part of her philosophy is to connect people to the land, and, to do this, she wanted to build a community around the farm. They built the one-room eco-cabins and a large community area complete with a communal fire pit.

While working on the cabins, they produced crops, began a farmers market, operated a CSA, and raised heritage turkeys. Money was still tight and they were looking to diversify; they started farm camps and events.

They staged a music festival the first year with the goal to let people know that the farm was there and what they were doing. By the fourth year, they hosted more events ranging from private gatherings to goat yoga. The yoga turned out to be a big hit. As it turns out, goat yoga is one of the more profitable enterprises. The farm camps are a personal passion for Fran.

Fran says you never know what is going to work for your market, so you just keep trying different things.

DEWAYNE HALL

Red Feather Farms – Tulsa, OK

When I met Dewayne, it didn't take long to see he was someone I wanted to work with; genuine people are hard to find but easy to spot when you meet one. That day, as I learned about him and his operation, I realized I wanted to interview him for this book.

Turns out, Dewayne is the oldest person I interviewed because that's how he started our talk. "I'm probably one of the older people you've interviewed for this, right?" He followed up by putting me on the spot and asking how old I thought he was. I guessed about 50 years old. He laughed, "I'm 60." We spent the next few hours discussing his life experiences.

Raised on a farm, Dewayne went to college to be a microbiologist. Computers were becoming the norm, and all testing was being done by computers. He didn't want to be stuck behind a computer, so he switched his degree to agriculture with just six hours remaining in his degree path. To me, that was a weird jump, but he explained agriculture is the precursor to nearly all medical research.

During the switch in degrees, he picked up his teaching hours and taught school for five years, finding a love of teaching. After his time in the classroom, Dewayne moved into financial services. When he found himself wanting to get back to his roots, he started teaching again. This time, he taught people about Victory Gardens and how to identify wild edible plants. This progressed into teaching herbal medicine.

He knew he needed to be living on the land because that's what made him happy, and he could have the most impact on the world, so he started looking for land. It took three years before he found the property he was looking for; it had been abandoned for a decade but had potential.

Dewayne feels we live in a throwaway world and that translates into the way we treat each other. Treating people better means treating the Earth with more respect and living in harmony and caring for one another as a worldwide community.

The first job on the property was to clean it up to make it functional and that took a year. They spent the time and effort asking Mother Nature what needed to be cleared and what needed to be left alone. One of Dewayne's core beliefs is living in harmony with the land to learn the will of the property; that's why they were slow to cut and clear to see what the land had to offer.

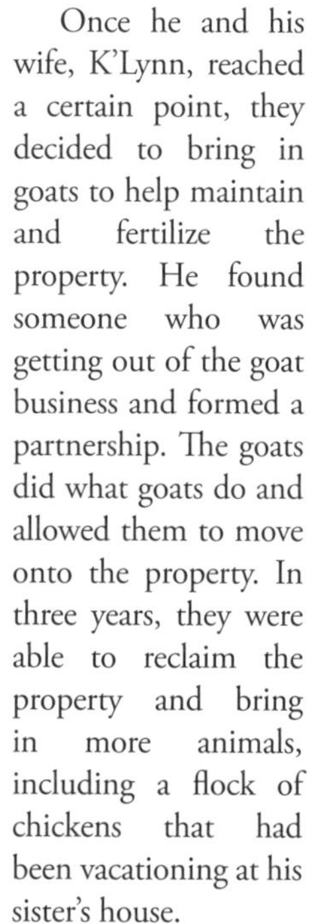

Once he and his wife, K'Lynn, reached a certain point, they decided to bring in goats to help maintain and fertilize the property. He found someone who was getting out of the goat business and formed a partnership. The goats did what goats do and allowed them to move onto the property. In three years, they were able to reclaim the property and bring in more animals, including a flock of chickens that had been vacationing at his sister's house.

During this time, Dewayne was teaching at other locations about what he was accomplishing on his own property. Dewayne knows a farm can't be 100% sustainable, but that was the goal: to make everything as self-sufficient as possible.

"I want to be able to have a classroom to teach, then walk out the door to have a hands-on experience," he says.

Dewayne sprinkled in his agritourism story throughout our conversation. He sees agritourism in three different forms: products and services, education, then entertainment. Being true to himself, Dewayne's main focus is education backed by a few products and services like elderberries, goats, chickens, eggs, and herbs. He says, "I want to change the world, and the only way to change the world is to change *my* world. My agritourism story is education ... teaching (people) how to live in harmony with what they're doing and not depending upon the chemical process of things."

MATT WILKINSON

Hard Cider Homestead–Ringoes, NJ

Matt W. grew up in central New Jersey and his farm story begins there. His maternal grandfather was from Italy and worked in New York City as a salesman. His grandfather's lifelong dream was to have a little farm of his own so he saved up to buy about 40 acres in a town called West Windsor, New Jersey. This farm is where Matt caught the agricultural bug.

Years later, he started saving money for his own farm, but he went a step further and earned a degree in agriculture. He eventually purchased a farm about 18 miles from where it all started, and he's been there for more than a decade.

By day, Matt is a public high school teacher and, in the early evenings and all summer long, he becomes a farmer. He has been an educator for more than 33 years and feels teaching and farming go hand-in-hand. After many years in the classroom, Matt has developed a gift of gab and an appreciation for speaking with people. He says being able to communicate is a major asset.

Matt W.'s agritourism story started when he was establishing his farm and felt it was a shame people didn't know about his place. He was going to the Mother Earth News Fairs and learning about new concepts and practices. Being a classroom teacher, he was trained on all the ways people learn. "Going to conferences and listening to speakers is great, but I knew there were a ton of people who like to see, touch, and smell," Matt says. "Sometimes, you just got to get dirty in the dirt."

That's when Matt decided to start with on-site interactive classes. The classes grew, and the number of students kept increasing. People kept coming in from farther and farther away until he had people from as far away as Ohio wanting to attend workshops, but, like most people, they didn't have the extra money to travel. That's when Matt started allowing visitors to pitch tents on his land, which turned into another venue to stay on his farm.

Matt put out the word by giving presentations and talking to people about what he was doing on his farm. By providing quality information in a hands-on experience, Matt was able to grow the agritourism leg of his farm, which continues to grow today.

JOHN MOODY

Some Small Farm – Irvington, KY

John's farm story started when he began running a buying club in Louisville, Kentucky. About 15 years ago, the club needed a supply of pork products, and John was having trouble sourcing the right amount. He was trying to obtain non-GMO meat or grass-fed animals. And since he couldn't convince local farmers to raise what he needed for his business, he decided to do it himself. John had other reasons for getting into agriculture; he always wanted to have land and all the benefits that come with it, but pork was the push for getting started.

John, his wife, Jessica, and their family started in agritourism by offering classes and workshops to fill a need for homestead education. When the opportunity to hold a session arises, he feels it's worth the time investment. John has found it comes down to figuring out what works for you and doing more of those elements.

Most of the classes at Some Small Farm center around compost and soil building with an introduction to planting. The class model was to get a group of people together and let them vote on each class's subject. The students would dictate what they wanted to cover with what was going on at the farm at that time.

Once or twice a year, they open the farm for an event. Some farm days are free, especially for groups of children, while they charge for other farm days. John points out the necessity of things like porta-potties, trash pickup, and infrastructure, all of which costs money. He says having 30 youngsters plus 10 to 30 parents is a lot of work, even if the visit is only for a couple of hours. It's best to have a value-added up-sell product associated with your farm day, if nothing else, to break even. He also recommends having items for children to take home after being at your farm.

JAMES HALVORSON

Halvorson's Hidden Harvest– Temple, TX

James started his farm journey with a kitchen garden that turned into a full-time business, something he never intended. This kitchen garden then became a passion for sustainable, homegrown food, which led to a total career shift. This endeavor has evolved over the last 12 years; he takes his knowledge to the community to let people know they don't have to inherit a large piece of land to start a farm. James says he and his wife, Amy, were like a lot of Americans—just getting by. At this point, they went out on a limb to sell their residence in town and bought a piece of land with an older homesite. They started working on it and built their new life out of the ashes.

His agritourism story grew with his passion. With a small start teaching on the farm, James aspires to be an educational platform where people can come out and follow every step of food production, learn where their food comes from, and even help make quality value-added food items.

James wants people to know you can start your own plants for less than 5 cents each and to realize gardening can be done inexpensively. Part of his gardening philosophy is the mastery of reusing, reducing, and recycling because everything can serve a second purpose.

He enjoys bringing people out to the farm to teach them all the steps of farming so they can leave with the confidence to replicate what they've learned when they get home, whether it's gardening by the square foot or by the acre. Another point he likes to make is you can support a family with enough vegetables in a 10-by-10-foot garden bed if you manage it properly and put in less than an hour a day.

He wants people to know how to preserve what they've harvested, how to prepare, and how to deal with leftovers from compost to chickens.

In James' agritourism model, he teaches, shows, and develops classes to help people learn to know how to do things on the land.

MATT STEPHENS

Matt Stephens Family Farm – Temple, TX

At the Matt Stephens Family Farm, we want to share our love of agriculture and the outdoors. While people are at our farm, we offer education during a fun experience. On our farm tours, I like to stop the tractor in the middle of a field to say something like, "Now that I have your attention…," then proceed to tell them what I'd like them to know.

When I set out to write this book, I knew I wanted it to be a collaboration with other agritourism farm owners. A single individual cannot know everything there is to know about any given subject. I say, "Surround yourself with good people heading in the same direction." This book is a community of people gathered to pass along what we've lived and learned, and what we continue to grow from daily. I asked myself and each owner a series of open-ended questions and the answers were beautiful. In fact, some took me in directions I never considered.

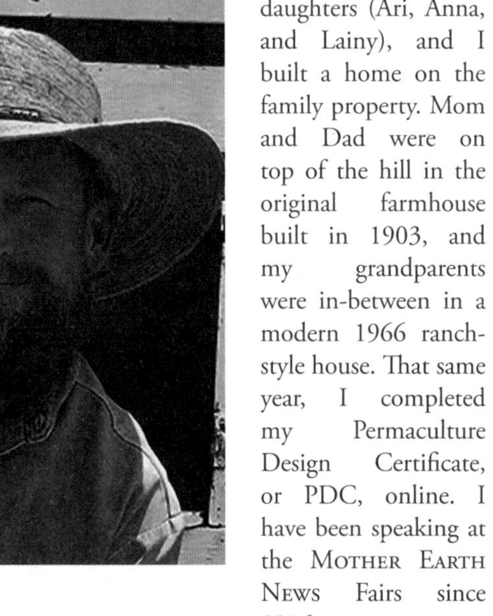

I want to unite this agritourism community in order to save the family farm and help small farms prosper again. There is no forward progress if everyone thinks and believes the same way. That's why it's vital that we talk and communicate with one another, sharing new ideas and lessons learned. We need to keep each other inspired, so we all know and believe it can be done.

I was born and raised on a central Texas hobby farm where I learned from parents, grandparents, and extended family about rural life and Mother Nature. This is where I developed a love for all things that live and come from an agricultural lifestyle. Our story started in 1961 when Grandpa, or Pap, decided to buy a run-down farm just north of Temple, Texas. My dad's dad grew up in the wake of the Great Depression and lived in a subsistence farming family most of his young life, so he knew how to farm to survive. By the mid-1980s, I was well on my way to learning how to farm and ranch, and to love Mother Nature.

Years went by with dirt under my fingernails and a passion growing in my heart for all things that come from the land, both domestic and wild. Eventually, I left the farm for life's adventures. The biggest adventure was getting married. As a young married couple, we migrated around Texas for schooling, work, and the sheer fun of new places. Like many people, we wound up back in my hometown and finally on the family farm.

In 2014, my wife, Summer, our three daughters (Ari, Anna, and Lainy), and I built a home on the family property. Mom and Dad were on top of the hill in the original farmhouse built in 1903, and my grandparents were in-between in a modern 1966 ranch-style house. That same year, I completed my Permaculture Design Certificate, or PDC, online. I have been speaking at the Mother Earth News Fairs since 2016.

Our agritourism story started well before we went public. Mom asked me for a diversion during one of our large Easter/family reunions, after the egg hunt and smashing confetti eggs. A lot of the youngsters were getting restless while the adults were enjoying visiting. I suggested we hook up the flatbed trailer and take a hayride. The trailer was small, only 5-by-10-feet, but it was big enough for the children who wanted to go. I pulled up, ready for an expedition, and hollered, "All aboard!" We left the yard to tour the farm. By the time we got back, the remaining children and half the adults wanted to see the farm from the trailer.

Fast forward a couple of years, and I was doing small private events on the weekends while teaching in an independent school district during the week. Co-workers discovered I was a farmer/rancher and started asking me about animals we had and what crops we raised. They also asked if we ever allowed visitors to see the operation, and I said yes. The comment kept coming up, “You should be open to the public.” It was said so often that we started taking a good look at doing public events as opposed to just private ones, and that’s when the real journey began.

We did our research on going public, and I worked for two local agritourism farms before we opened our property to the public, all while continuing to hold small private events at our place. Working for other operations to gain on-the-job training was crucial for insight into what I wanted in my own agritourism operation and what I could do without.

We are a working farm and, by design, agritourism is one of the legs that supports the whole operation. We started hosting a few public events for small groups to test the waters. We bring people in for fun and tell them: “To save the real food systems, to grow natural food, grow real people, save the family/small farm to save the world.”

What is my purpose? I want to inspire you to dream, educate, and warn you with my story. This book is an exploration of my years of working towards and dreaming about bringing people to our farm to share my love for natural farming and farm fun.

SECTION

One

GETTING STARTED

Growing in agritourism through research and planning.

CHAPTER 1: CONSIDERATIONS FOR BEGINNING AGRITOURISM

If you believe and have an abundance mindset, it can happen for you.

This book is going to ask questions to prompt answers that will relate personally to your situation. Some of the answers will be encouraging and some not so much, but they can be your guide to a great life with agritourism.

The six other farmers featured in these pages amaze me with our similarities in the areas discussed. Under the banner of agritourism, we came together with our individual fields of expertise and our different viewpoints, and we have a lot of the same things to say about the fundamentals of what it takes to be successful in agri-business and agritourism. Our experiences are packaged together and placed in your hands so people can avoid making the same mistakes and can share in similar successes. Learn more about these farmers beginning on page 10.

I feel it is important to start with a growth mindset. A growth mindset uses the old adage that problems and setbacks are stepping-stones rather than stumbling blocks. This way of thinking looks at challenges and failures not as knocks against us but as progress leading to future successes. You *must* know you *can* make it happen.

It may sound cliché, but you have to *believe* you can be successful before you can *be* successful.

Another guiding factor I use is to have an abundant mindset. My definition of this type of mindset as it pertains to agritourism is that there is room for all. If we are true to our hearts and follow our own directions, there will be space for all our endeavors in agritourism. You can have two peach farmers next to each other, and one likes U-pick and the other likes to make value-added peach products. This allows both individuals to express their own wants and desires for their operations and be successful. Find yourself in your enterprise. Truth be told, the two farms in this scenario could actually promote each other for both to be more successful.

Whether you have a current operation and want to add agritourism, or you want to start a new agritourism business, you have to know what is in your heart. Farmers and people of the land are generally passionate about what they do. To include agritourism in the mix, you need to be specific as to your personal intent. You can plan and design a lot of the aspects of your operation, but you still have to love what you're doing.

This chapter takes a look at your personal plan, your business plan, and your possible target market. We'll delve into more details in later chapters.

Your Personal Plan

To know yourself solidifies your goals and dreams. Honesty with yourself is key to being successful in agritourism. This honesty will help drive the design of your operation. Asking yourself specific questions up front, especially hard questions, is better than being blindsided, or worse, losing everything you have worked to build.

One factor of being successful is knowing your dreams and how big of an operation you eventually want to own. For example, if you simply want to teach a few people each year about backyard chickens or backyard greenhouses, then a few hens or some tomato plants could get you where you want to be. But if your goal is to be a regional influencer for sustainable agriculture, you're going to have to dive deep into what it will take to be that person. Research will benefit you in both endeavors, saving time with fewer headaches along the way.

Look internally to figure out your personal gifts and interests.

You can't imitate another operation and truly be successful. It doesn't matter how prosperous the other farm appears to be, they may be barely getting by or it simply may not work with your strengths. Your passions and strengths will help you build your personal success.

I once tried to copy a local agritourism operation and almost lost the farm. It was simple, they had a pumpkin patch and I never had one. I thought I could just add pumpkins and everything would be great. This was flawed thinking. The weather turned and my pasture roads were flooded. I could not open for pumpkin season that year. So I had to wholesale my load of pumpkins to my neighbor who was located on higher ground. It was not a break-even situation, but I learned a lesson: Mother Nature rules.

What can you produce on your property? If you want to raise grass-fed bison and you're inside city limits, your location probably doesn't fit the desired

IMPORTANT QUESTIONS TO ASK YOURSELF

How does your farm plan fit into your personal life?

How does your farm dream affect the other people in your life?

What direction does this plan take you?

Do you have the heart for agritourism?

product. You would be better off running a pick-your-own market garden than any type of livestock. When it comes down to discovering what your product can be, decide on one while still being willing to try different options. You may decide you want to feed the world or teach the world how to feed itself. Then you may draw the line at not growing anything ornamental and find the beauty in food that comes from the land. Then again, somebody may have a field of bluebonnet flowers and invite people out to enjoy the beauty and time with Mother Nature, and they may decide that's how they're going to contribute to the world. Both are contributing to a better world, just from different angles.

To know when you have "made it," you need to define what success means to you.

If you asked a thousand people what success was to them, you may get 999 different answers. In the corporate world, success is scored by a balance sheet. In this, no one else's opinion matters, only yours. You might want to provide for your family and make a difference in your neighborhood. Your efforts might be worthy if they make a positive impact in your city.

We have to know our own personal definition of success to discover when we have accomplished our mission. Without this simple definition, we may get lost.

Are you a people person or are you a hermit?

Do you want to live in the middle of your property and not see anybody? If you vote to be a hermit, then agritourism might not be for you. You can carefully design your operation if you are not the most outgoing, but you still need some desire to connect with people. One way to test your people skills is to imagine yourself striking up a conversation at the grocery store if you see someone buying carrots and you are a carrot farmer. This can give you some insight into how comfortable you are with people on your homestead. How much of a people person are you?

To make your business work, master being you, being authentic. People may be bold enough to open your closed gate, but those same people can spot a fraud. In this scenario, if you talk to them while gritting your teeth, they will figure customers generally make you mad. But if you tell them the truth that their off-hours visits keep you from opening on time, they will know you are truly working for them. Be authentic.

Most farmers do what they do for the love and joy of farming.

No one should work as hard as a farmer or rancher unless they absolutely love what they do. This includes an agritourism business. When you deal directly with Mother Nature and the general public, you deserve a medal. But if you love what you do, it isn't work. You have to be passionate about what you're doing, whether you're raising crops, working with animals in extreme temperatures, or just fixing the fences.

The endless hours we put in need to be fulfilling to us as individuals or we'll burn out. Burn-out happens when you do not enjoy your labors or, at least, the end results of opening day.

Look for a community.

Agritourism can be lonely. Not everybody understands your business. That's one reason for this book. We do not have to be alone; there are others who understand what you want to accomplish, and they want you to be successful.

Entrepreneurship is a solitary venture. Even if you have a partner who shares your vision, then it's still two people. Farming can be just as reclusive, but outside people understand most farming practices and support your endeavors. It's my belief we need a support system of people with experience in agritourism to call on when we need help. This type of community understands what we're trying to accomplish.

When I take the stage or talk to any size group, I make time for the audience to talk about their operation or dream operation to the person next to them. I modified this technique after watching Matt Wilkinson speak. Since then, we have conversed about our agritourism operations and created podcasts together, and those were major factors as to why I interviewed him for this book. I feel it's important to build community wherever we go; it's a win-win.

Sometimes just knowing there is another person in the universe who understands your dream is enough to get you through a bad day. But on the really tough issues, being able to call an agro-friend could save the farm. I want this book to help build a global community around agritourism. Our operations may be miles or oceans apart, but we can stay connected. I consult with people all over, and there are more similarities than differences when it comes to agritourism operations.

Can you remain adaptable and resilient?

In business in general, and especially in agritourism, you have to remain adaptable—a core characteristic of anyone in agri-business. Things will not always go as planned, and we need to be okay with that for the best possible outcome. We need to expect setbacks and be able to adjust when things do not go the way we planned. Know that you will overcome whatever comes your way.

Along with being adaptable, being a resilient, lifelong learner will help you stay on track. This constant research mode will keep you studying new ideas, new trends, and better ways of operating. This mindset also helps you figure out what people are looking for and what experiences you can give them on your farm.

One of the examples given to me was about a gentleman who was making more than $70,000 a year and could not understand why he was not making more money. The reality was one had to drive 30 miles to get to his place and then maneuver down a long, muddy pasture trail to a worn-out corn maze in the lower part of a field, which was waterlogged most days. And that was the extent of the fun. When the new owners took over, they expanded to meet the wants of the people. They added pumpkin carving, basket weaving, barrel trains, hayrides, and the usual festival attractions. This, in turn, translated into overwhelming financial success. Stay adaptable and resilient.

Your Business Plan

After you look inward to find why you want to run an agritourism business, you need to look outward for the business details.

A business plan is a way to design your operation; it's a road map for you to control the direction and destination of where you're headed. Planning your business means you'll be planning your work. To keep you on the path to success, business plans need to be written down as a living document, which is fluid in nature and not a strict contract that can't be changed. This document can be as short as a single page or as long as hundreds of pages. I prefer the shorter version to keep things simple, but it needs to be long enough to contain all the essential materials to help keep you on the side of success.

Start small and grow from what works, learn from what doesn't work, and don't add to your operation too fast. As part of this planning research, you also need to answer the question: Is this leg of your business self-funded or will you need to get outside funding?

Using a business plan is essential for a solid base, and it will outline what is reality and what is your hope for the future. Risk is associated with being in business, and a proper plan will help you weigh the risk—especially when you plan to leverage anything. Taking out a loan or borrowing in any form needs to be extensively evaluated. It is not advised to leverage against an agriculture venture including agritourism.

Markets change constantly and this could put your farm at risk. For instance, if you want to add short-term rental cabins to your current operations, it would be better to plan to build one cabin, and then build over time to the number you originally planned by using your profits as opposed to taking out a loan for the full amount. Do not borrow more than you can financially lose.

In a proper business plan, there will be a strength and weakness analysis.

This will include your interest/aptitude to know what tasks you should do and what you should hire out. If you love building with your hands and are good at carpentry, you might not need to spend extra money on this type of work. Doing the work yourself, especially in the beginning, can save you cash because labor is among the largest expenses in agritourism.

It's important to know your own strengths and weaknesses. You will need to be a jack-of-all-trades, from growing plants and raising animals to managing employees and customers to keeping the books and creating a marketing plan. Marketing is covered in detail in a later chapter.

When people visit your farm, they are there for an experience.

In developing an accurate plan, you need to know your target audience, which will lead you to what they want and are willing to pay to experience. And with each experience, you'll need to know the potential add-ons or upsells that your patrons may like or not even know you offer, such as how-to classes, fresh eggs, healthful produce, or bicycle tours. Do not expect to meet every possible scenario but be diversified to accommodate different tastes. We'll discuss narrowing down your target audience later in the book.

What are the food and beverage requirements for your operation?

Besides acquiring the wholesale items required to serve food and drinks, what kind of license and permits do you need? Such as food handler cards for serving or vendor permits for cities and counties.

Every branch of agritourism has its own set of regulations, industry standards, and specific safety guidelines. For example, liability for a petting zoo vs. insurance for youngsters riding horses or other animals. Some of these rules change annually so it's important to keep up-to-date with your specific industry or industries if your operations overlap.

What kind of buildings will you need?

If you have a location or are planning on acquiring one, the questions are the same. Analyze your needs for infrastructure: food, water, shelter, and energy. Two advanced considerations will be access to your location and let's not forget toilets.

Are temporary tents a possibility? Will you need to renovate an existing structure? Is it time for a new construction project? Where is the energy or power coming from? You can live off-grid pretty comfortably, but a commercial operation is going to take some careful consideration to keep the public happy. Once people decide to visit your farm, how are they going to access your property? Don't expect people to come back if they get stuck on a muddy field road. People need to be able to come and go with ease. How are you going to let everyone go to the bathroom? Is your septic system large enough to handle public use or are you going to bring in portable restrooms? Renting portable toilets can be costly, and you will need to evaluate the expense of having blue boxes sitting at your place.

Examine Overhead and Finances

Overhead is the cost of doing business and needs to be considered for each venture of your agri-business enterprise. As an example, you have a short-term rental cabin, a sheep herd to help you teach small-scale shepherding, and a pick-your-own strawberry patch. If one section is not profitable, then changes to the total business need to be considered.

In any enterprise, finances should be at the top of your list for discussion with your business partners and/or spouse. Do you have another person you are responsible to for your business activities? If you have a spouse or business partner, then communication is key. My wife and I did not start out meeting about farm business regularly. Without this regular communication, I started taking the business in a different direction, and that's when we almost lost the farm. She reminded me that our core goals for our operation were education and fitness. From that day on, we set aside an hour every Sunday afternoon to discuss the past week's accomplishments and the next week's goals—for both of us. As our operations have grown, we now meet daily for about 15 minutes over breakfast.

Whether you are open year-round or seasonally, you will have proportional costs associated with how long your gates are open, and you will have to generate enough revenue to cover the added infrastructure. The extra open hours add to the wear and tear on your farm and things need to be maintained, including access to your farm. The road is critical because this is the first and last thing people see of your operation. All of these are considered part of overhead.

from Dewayne Hall, Red Feather Farms – Tulsa, OK

If you plan to operate year-round, then there will be heavy and light times each year. Estimate what your maximum number of patrons will be during both times. Then evaluate drop-off and pick-up costs and monthly rental charges. As an example, for seasonal or heavy traffic times of operations, weekend or weekly charges are usually not cost-effective and you will need to evaluate them for your specific event. For your season, you may plan to be open for six weeks, but the restrooms would sit empty for two weeks. If the port-a-potties have a monthly charge, the extra two weeks could take all your profit. This is where you need to calculate whether extending the season two weeks or cutting two weeks will be more cost-effective.

Managing the public can be a hidden expense. Even if you have operating hours posted, people will show up at all times and, if the gate is open, they will come through. An open gate or unlocked gate policy could cost you time, energy, and money to deal with these random folks who don't know better. My farm is closed to the public and events are strictly scheduled to limit public access. This allows me to work my plan and stay focused on what needs to be done. In this plan, you need to honestly evaluate your resources; think about your time, available capital, your land, equipment, and

from Fran Tacy, Franny's Farm—Asheville, NC

all the assets you bring to the venture. This is where it's better to start small and rent a tractor twice a year to get the work accomplished rather than take on debt for a $20,000 piece of equipment that might sit idle most of the time.

Insurance is another cost that will be covered later, but for now, consider if your insurance costs $600 a month and you're grossing $200 per weekend (an average four-weekend month) then you're only making $200 to cover all other expenses.

Target Market

Is your location close enough to allow your target audience to drive to your location, or can you afford to transport the goods to the people? Narrow down who you will focus on to sell your land's bounty. Most people are too vague in answering this question. Something like "Everyone who eats meat" is too broad of a category.

The word *demographic* often has a negative connotation in the media, but, in business, it's vital to success. We have to hone the definition of the group we want to serve. This helps in your marketing plans, and marketing is how we get folks to the farm. In the above example, simply targeting all meat eaters will waste a lot of time and money if your farm produces all natural grass-raised beef. You are not raising just any old cow. You are producing a high-end product that not every meat eater will appreciate, much less spend the extra money to acquire.

If you are operating a pumpkin patch, the same principle applies. You want a clear vision of your customer so your marketing message will speak directly to them. The target market is not everyone who likes fun on the farm, but more likely young families with children from the ages 5 to 12. Yes, other people will show up too, but by having a laser focus, you cut out the wasted advertising dollars.

Your venue has to attract people so they leave their comfort zone and spend money with you. Some people will not drive across town to select a Christmas tree. Other folks will drive two hours to pick the perfect centerpiece for their holiday tradition because that trip represents a day-long excursion to a winter wonderland of fun and games.

For the reverse—getting the product to the customer—you'll need to consider transportation costs, which is part of your overhead.

A dangerous mindset of agritourism is when people become stuck in what they want to do and are not willing to change or try something new. When this happens, we forget we are dealing with our target market and that they are the ones who will tell us exactly what they want or need. There must be a balance between what the customers expect and the work you want to do.

NOTES

from Dewayne Hall, Red Feather Farms – Tulsa, OK

CHAPTER 2: IS AGRITOURISM RIGHT FOR YOUR FARM?

More importantly, is agritourism right for you?

Most people are uncomfortable talking about themselves and their feelings. So if you find yourself wanting to skip this chapter, you probably should read it twice. In my talks about agritourism, one of the first things we discuss is what's in your heart. That's one of the hardest things to discover. This chapter will help you find your purpose. After talking to many successful operations, it's interesting that most mention *being true to yourself* when it comes to business. No one goes into agri-business without it being some inner part of themselves.

In business, especially agri-business, you'll be marketing yourself, your skills, and your property. You are adding value to the raw products produced on-site, and you have to attract people who are interested in what you offer. It's a benefit to be the type of person who connects with your community and who enjoys attending Chamber of Commerce meetings and other community events. An outgoing personality helps you build your brand. Be inventive and resourceful with your land as well as with your ability to network. Remember, you cannot be everything to everyone. So be honest with yourself when determining what you want.

Be realistic about *you*.

If you do not like getting your hands dirty, then I suggest you avoid a market garden. An agri-business has to be enjoyable for you in order to work. You don't want to dread opening the gate every day. If that becomes the case, your customers will feel it and they won't come back, or you won't keep doing it. Be realistic with the number of hours you want to work. This will be key in the design phase of your operation. If you want a 9-to-5 operation, anything agri- is not going to fit your plan. If your personality type is such that you need a steady paycheck, then most agriculture businesses, including agritourism, are not going to be your answer. With that being said, you can design your agritourism to fit a lot of your desires.

Have a clear definition of your personal "why."

What is your purpose for doing what you are doing? Why are you choosing this particular area of agritourism? Find your motivation. Having a niche market is a benefit because it makes you unique. In the beginning, you'll probably waver some. You'll try a few ideas before landing on your true niche. Whatever you decide is your "IT," the one thing that is yours, be the best at it. This doesn't mean you and your business won't evolve over time; this is natural and should be evaluated annually.

As an example of an evolving operation, before I started raising animals on my farm, I did my research into species, breeds, characteristics, and feasibility. This led me to American Blackbelly hair sheep. As the agritourism grew on our farm, we learned people preferred Painted Desert sheep; they have all kinds of pretty colors and splotched patterns that make them more fun and interesting. We are now breeding for our feeding zoo and are selecting for different colors and patterns because that's what visitors want. We still keep a few American Blackbelly in the herd, but when it comes down to it, a hair sheep flock is what I wanted for my operation because of other characteristics. The detail of what they looked like was less important; even though I started out wanting one thing, I was able to be adaptable and transition my business.

Partner with someone who is grounded, particularly if you are a dreamer with grand ideas.

Whether it's a business partner or spouse, make sure they are rooted. If you are a grounded individual, then I suggest discussing your business plans with someone who sees the world in a grand way. At least buy them a cup of coffee every now and then to generate new ideas for your business direction.

In my life, my wife, Summer, is my grounding force because I am a huge dreamer. I can plan it all out for 10 years and know in my heart that it's going to be amazing. She reels in my ideas to a more manageable size and reminds me of the original goal. One specific time was about chickens. Did you know there is a big difference between 10 chickens and a hundred chickens? My dreamer mentality goes for the hundred chickens every time, but with the help of my wife, I'll start with 10 chickens and grow into the hundred, if that's the direction we're going to take the farm.

If you are married or have a significant other, you need to take them and their personality into account. Opposites usually attract, so how does this fit into your business design? If they are more introverted,

IS AGRITOURISM RIGHT FOR YOU?

they may not want to be the face of the farm and interact with the public, but they could help build the "sections" on the farm or manage the animals and garden. You can be the face of a successful pick-your-own operation and have employees or other family members handling the customers on a day-to-day basis.

For family farms with multiple farmers and possibly multiple generations to handle operations, it's important to have everybody in a role that best aligns with their personality. Introverts may want to stay in the office and handle the background details, while more outgoing family members might want to be in the middle of the chaos that is a farm festival.

Some agritourism operations are more people-intensive than others, and, by design, you can accommodate different personalities.

Agritourism is a people business.

When you're in the people business, situations may not be ideal and people may be frustrating. Whether it's the weather or somebody got in an argument during the car ride to your farm, *people are people*. Can you deal with customer complaints? Customer service is important because sometimes when they come to the farm, people just don't know what they're getting into, and this is a great opportunity to teach them. The more you can help them, the better it will be for the customer.

Also, the people who come out to your farm are generally not going to care about your property as much as you do. You know exactly what it took to put that fence post in or the energy it took to make the maze or the building that holds your products. So when they hang on the gate or damage property just to get the perfect photograph, does your personality let you handle those situations with grace?

During our conversation, James Halvorson of Halvorson's Hidden Harvest had somebody show up at his house unexpectedly, a perfect example of the need to be ready for people to show up at any time. You need good fences and a good gate with visiting hours posted. This is not just for the short-term rentals, it's when you open your farm to the public for any reason. Just because your hours are posted as being 8 to 5 Monday through Friday, it won't stop people from showing up whenever they want. Even if your voice message and your website have it spelled out, people will try. I had a friend operating a pumpkin patch and had people climb over a barbed-wire fence to take photographs in the pumpkin patch after-hours.

I have found that most people will be respectful and 99% of people will do the right thing.

However, there is always that 1% who won't be. A certain percentage of time, money, and energy will have to go to repairing and replacing damages from agritourism patrons. Plan for that small percentage of people who think the rules do not apply to them.

Even though you have the rules and guidelines marked and posted clearly, will you be able to give mercy and not rip their heads off?

If your personality is such that you are going to stay angry over the fact that you had to replace and repair the same hinge 10 times, you may want to

look at a redesign. It could be simply doing without a gate at that spot. Be honest with yourself about what you're really trying to accomplish when designing your operation, and it will save you lots of headaches and heartache.

Set boundaries with yourself and others.

Design limits into your business. Some you will see right away and include them in your original business plan. Others will evolve over time as your operation grows. Consider your stage of life. Life is complicated enough and adding agritourism could make it more entertaining or more chaotic. For instance, if your family unit is in the childbearing years or y'all are looking at adopting, it's a good idea to design your agritourism operation around your agriculture season. If you have a pick-your-own apple orchard, don't plan any life-changing events during apple harvest. Consider family birthday parties or anniversaries when setting boundaries for the farm. You can't control everything, but with a little forethought, you can mitigate some of the headaches. Making good plans can help you navigate some of these life issues.

Keeping the neighbors happy is another consideration.

If you have a thousand cars a day stopping by your farm, is that a hindrance to your neighbors? If it's a challenge, could there be other avenues designed that alleviate those conflicts? Can you obtain an easement to access your property from a different road with no houses on it? Sometimes thinking outside of the box for new solutions is crucial to keeping the peace with those around you.

You have to have the ability to plan your work, work your plan, and stick to it.

If your income depends on large seasonal events, planning a four-season operation can be crucial to your success. Depending on how you design your operation, you may never be able to leave the farm. If you are going to be open to the public, that means you are there 24 hours, 7 days a week, for 365 days. You can hire somebody to do the work and manage the operation, but you have to be intentional in your design to be able to leave the farm. Your end result will depend on your design. However, by design, you can eliminate some of these constraints. One misconception is that you will always be able to find somebody to do the work for you. No one is going to care about your operation as much as you do.

Even the most extroverted person will need a break from the outside world. If these boundaries are not in place, you will eventually feel the sting of burn-out and this is a dangerous place. You can lose your passion and your farm if you don't take care of yourself. You may have the most unique and wonderful operation, but the farmer has to stay healthy just like all the plants, livestock, and guests.

The competition will grow, and if you have a good game plan, you can outlast the competition. Most of your competition will not have a solid plan, and they will fade away. As long as your design is solid and you're doing right by your customers, you will survive.

Think about Your Land

It is my belief that agritourism is for everybody who loves anything related to agriculture, and by my definition, agritourism is for anyone with a dream. There are so many types of agritourism that it could work for almost any farm. John Moody of Some Small Farm says, "In my opinion, every farm should offer some type of agritourism."

Again, some types of agritourism may not fit your particular farm business model. Even if you do not have a farm—yet—you still have options. Not only do you have to be honest with yourself, you have to be honest about your property.

Ask yourself if the current design of your agritourism operation is right for your land. If you own 160 acres of irrigated cotton, it may not be the right location for a pumpkin patch. Then again, it might be perfect if other demographics line up. Do the research for what is needed for individual crops. You can't have a sustainable grass-fed bison operation on 5 acres. The same could be said for a 50-section cattle ranch in deep west Texas. However, if you have it in your heart to run a dude ranch and are able to drive to the airport to pick up your clientele, it could work.

You can develop some kind of agritourism for almost any property. You simply need to line up your passion with the land.

When it comes to your property, one of the first things you need to think about is access.

How are people going to get to your farm and then once they're there, how are they going to move around your event? It's okay to use a worn-out pasture road to get the chores done, but customers should not have to go down a cow path to get to your property. The majority of people won't do that and your operation will suffer. To go along with the cow trail, how is Mother Nature going to affect the accessibility of your operation? Is a little bad weather going to shut down the whole thing? It's something to consider when you want 5,000 cars traveling down your driveway each day. Is your infrastructure going to handle the workload you're designing? You can start small with chickens or rabbits and evolve your idea from one location to the next. If your dream operation is bigger than the property you're currently on, you need to be okay with moving to a new location when needed.

You could partner with someone, but be careful with this option.

If you have a full-time job off the farm and are strapped for time, you can probably hire someone to manage any type of operation. Just consider what happens if you transition full-time to the farm.

Agribusiness people are resourceful. Necessary resources may be scarce, but you can make up for that lack. The barter system can still be useful in outsourcing some aspects of your farm. If you really do not like marketing, contract it out. We'll talk more about marketing your farm business in a later chapter.

Bonnie Chapa of Laughing Llama Farm tells how they developed into their property. She says: "The farm that you are on may not be right for you. Your dream operation may need a different location or you might have to build up to the ideal situation. In my area, the land can support seven to eight llamas per acre, but the question to ask is, 'Do you really need 80 Camelidae on-site to fulfill your dream?' Look at the other animals that may be on the property. We looked at the pastures we had and looked at what the grazing could become in the future. This brought up pasture management and what our fields looked like year-round.

"We had a pond with a spillway that stayed wet six months of the year. These conditions were just right to produce snails, which lead to liver flukes in llamas. Then there was the infrastructure. Which buildings had utilities: water, electricity, and sewer? And could we get the necessities for those structures?

"Our primary thought had to be given to access, how are we on a daily basis and how are patrons during events going to get around the land? Our big dream was right for this location, but we had to thoroughly think through all aspects of our mission."

Bonnie went on to tell about the fencing that needed attention before animals could be added to their new home. She also recommends looking beyond boundary fences. They had a 300-acre cornfield at a higher elevation above their property, and runoff from the field fed into a ravine that emptied into their pond, which could be problematic for the livestock.

You won't figure out all the details until you go through at least one cycle of the seasons and see how it all works together. Success can happen when you try to understand your little spot on Earth and see how it can evolve into your agritourism dream.

Any farm near a population of people has a good chance to succeed.

have access, you can make almost any property work for agritourism.

I found patrons will be forgiving if it's a unique experience, but everyone has limits when it comes to waiting in line with fussy children. So if they have to wait for too long to get to your event, they will find something else to do. You have to be organized, compassionate, and kind.

Being open to growth is important because you'll have to diversify to stay alive in this business. Once people think you're going to be successful and make a little bit of money, then everybody else is going to try it, too. Nothing becomes reality unless someone dreams it first. Look at the big picture to see what you're really dreaming. When it comes to your big picture, you have to stick to the road map you set or you might lose your way.

If some of these characteristics don't sound like you, maybe you need to find a partner who gets excited about this description or possibly pick a different industry.

honey extraction, Dewayne Hall, Red Feather Farms – Tulsa, OK

It's good to have a niche in the market with the understanding that a diverse farm is a stable farm. Most farmers who understand diversity quickly apply this concept to their land and will need to apply it to their business offerings. The wider your appeal, the more attention you will attract.

Also consider these aspects of your operation.

People will drive hours to see your wonderful venue, something they cannot see anywhere else, but will they have to get their shoes muddy? If you

As your operation expands and unfolds into what it's going to become, loosen the reins a bit. That way you can express some of your creativity and try new things to grow. You have to be in control of your ideas and dreams to be able to think them out thoroughly. And your dream will start to naturally evolve into something beautiful. Whereas if you were to start willy-nilly, your dream is likely to take over your resources and put you out of business.

NOTES

CHAPTER 3: DOES YOUR LOCATION MATTER?

The three most important rules of business are location, location, location — or are they?

Traditional business books will say location is key for a successful business. In many ways, agritourism is changing this old way of thinking. Many factors go into why location becomes almost irrelevant with agritourism.

Depending on your market, customers may travel hundreds of miles to get to your operation. Having a specific niche in a market can make your farm stand out from everyone else. When you have general competition in your region, narrowing your appeal can make your endeavor a one of a kind. If your position seems too neutral, look at your specific setup and find a regional feature that singles you out. Do you have a great site for bluebonnets to thrive or another species of wildflower that grows in your climate? Is your operation centrally located between metropolitan areas? This midpoint could be a focus in your marketing as the place to meet friends and family halfway.

As a general rule, location doesn't matter. Your locale will guide you in what type of operation you can have. I think anybody anywhere can operate an agritourism business or add a leg to their farm business. With a broader definition of agritourism, there's a growing trend for the farm to go to the consumer. In taking the farm off-site, it's a good idea to have the off-site farm event generate traffic back to your home base.

Some of the farmers I interviewed took a more rigid and traditional stance on business location. We agreed in many ways that business decisions will be guided by one's location; for instance, one would struggle to have a penguin petting zoo in Arizona. However, any farming venture can adapt to an agritourism component.

Success can happen with some careful planning and a well-thought-out design.

The farther away from your audience, the more unique the experience will need to be. If your operation is up in the mountains, and people are looking for remote wilderness skills, then your location is perfect. On the other hand, if your clientele is looking for simple classes on farming with worms, an inner-city site would be fine.

One of the farmers I spoke with for this book drives three hours to pick a basket of fresh peaches because three hours away the peaches ripen sooner. But where he grew up, his parents wouldn't drive 10 minutes down the road to pick strawberries because it was too far.

In Central Texas, things are more spread out. I've been known to drive two or three hours to take my family for a day-long road trip. We've even taken weekend trips, from Friday afternoon to Sunday evening, six hours away. In some locations, where everything is close together, this would probably be unthinkable.

To complement location, play by the rules and regulations of your property.

For instance, if you are planning a Christmas festival with carriage rides or hayrides, will those rides stay on your property? You live in the county; are there county ordinances against taking a trailer loaded with customers out on a county road? I'm not suggesting taking patrons onto a busy street just because there are no ordinances against it. Common sense must play a factor.

If your property is in the county, play by county rules. Are you outlining a big event on your place? What are the regulations for your site? What are the ordinances to have such an event? Are you selling hot chocolate or corn on the cob in your pasture? What are the food regulations for your locale? During your event, at what point do you need security? Will you have to get private security or county security?

When your operation is in the city, then play by the city's rules. Urban agritourism can be successful because people are able to walk to your operation. However, you're not going to find Berry's Bison Ranch near downtown.

In your agritourism dreams, do you want to teach people about raising backyard chickens? Do your city ordinances allow chickens, or are you going to need to partner with someone who lives in an area that allows poultry and establish your dream from there? In this example, your location matters, corresponding with your dream.

Take my farm, for example. The first half of the property is in the city limits and the other half is in the county. This helps me make decisions on what part of my operation happens on which half.

Don't forget to do your homework.

Bonnie Chapa really did her homework and researched the deed restrictions on her property.

Some deeds dictate what type of buildings can be built and where they can be located. For example, on my farm, everything but the house is portable. Not only do I want to rearrange all my chess pieces to where they fit best as I evolve my business, I also want to minimize my taxes. Taxes are a big deal, especially the difference between city taxes and county taxes.

Be even more cautious about doing anything on rented or leased property. Never do anything on property that you don't own outright unless you have an iron-clad agreement or contract that spells out every detail.

Do not kill your dream by not doing your homework on what is allowed at your location.

Is your agritourism supported by a localized population or are you looking to reach a four-state area?

In my area, we're close to a large military base that draws its personnel from all over the world. When you draw such a diverse group of people to your farm, it better be an extraordinary experience. People often are willing to pay a higher price to participate in a unique experience that is out of their way. By being inconvenient, that raises the stakes on how special the visit will need to be. Say you're the first in your area to have a memorable agritourism center, this adds to your uniqueness and the desirability of that experience. In our area, people travel two and a half hours to get to our local Christmas tree farm because it's that phenomenal.

Keep asking yourself, "What does my operation offer?" As the business changes, keep asking this question to stay viable.

As you look into the future, location may become more of a factor.

Location can be considered a marker of potential. As your farm dream matures, are you going to need to be closer to a certain population, or, on the flip side, are you going to need to be farther away from that same population?

Consider the cost of the land. If you decide to open a pick-your-own peach orchard, you'll need to be far enough away from town that your land costs will be low, yet close enough to a population center with people willing to drive.

James Halvorson says, "A guy once told me you can't sell river rafting rides when your land is in the middle of the desert."

Bonnie Chapa says location can be akin to accessibility. Usually, the farther out you get, the fewer amenities you have when it comes to road conditions. Her Laughing Llama Farm is located less than 30 minutes away from a nationally known hub in her market. This gives Bonnie a location of importance and proximity.

Before you purchase for your operation, ask these questions and seriously plan through your goals. It will save you some headaches and heartache.

How will people find you?

If everyone in the county drives by your farm on a daily basis, then people will know where you are. If not, your marketing plan will be crucial, so know your audience. Fran Tacy of Franny's Farm gave this example: There's even agritourism in the Andes Mountains; however, while the agritourism is there, it doesn't do as well as the agritourism in the cities of Peru.

NOTES

SECTION

Two

AREAS AND TYPES

Agritourism is a broad topic with endless variations.

CHAPTER 4: DETERMINE THE RIGHT TYPE OF AGRITOURISM

There's overlap in the areas and types of agritourism, which is where you can add value.

What are the areas and types of agritourism? A few are: pumpkin patch, Christmas tree farm, game ranch, pick-your-own fruit/vegetables/flowers, rural bed and breakfast, farm stay ... You get the idea. How to actually monetize each of these will be covered in later chapters.

To Begin

When you look at a subject as large as agritourism, you need to have convenient ways of breaking down the topic into more manageable parts, such as value-added items. Value-added will be covered in later chapters, but we'll touch on it here to show the diversity of agritourism. You can start a singular endeavor and mature it into something more well-rounded and stable.

Start with what items you raise right now that might generate sales. Say you are raising alpacas, you could just sell the fiber. Better yet, what kind of crafts and items could be made from the fiber? If you had a peach orchard, you could sell pre-picked fruit or give your visitors a better experience by letting them pick their own.

With a little creativity, the topic of selling items can be endless. You're giving the customer what they expect and providing them something memorable to take with them. In the case of Laughing Llama Farm, Bonnie Chapa sells the Dorper sheep that the llamas guard, garden produce, and, as a unique item, llama manure. During Bonnie's research, she discovered there is a market for nitrogen-rich llama manure.

One of the biggest draws for agritourism is the experience. People come for your produce, such as pumpkins, but they stay for the hayride, the petting zoo, the food, and the sheer joy of being at your farm.

Look at All the Categories

If we were to divide by overall categories, we might look at entertainment programs, educational operations, and product-based with value-added experiences. The possibilities are endless when you innovate and keep studying trends and events that draw in people. Successful businesses get people interested and excited to come to your farm.

Operations can also be broken into indoor and outdoor activities.

For instance, horseback riding or ATV riding, butterfly trails, bird watching, fishing, or hunting are all possibilities. Offering a spot on your place for people to take formal and informal photographs provides a unique setting and allows people to take a little bit of your farm with them when they return home. If you're set up for it, you could host concerts or events.

Having a farm day opens up new possibilities because there are aspects of your work you've been doing for years that people are interested in. A day with you allows people to explore and ask questions about how your dream works. A quick survey or simply asking people what they are interested in could provide you with valuable insights to adjust the direction of your agritourism.

One area of agritourism some people may not consider is game ranching.

It is similar to a typical farm and ranch operation. Game ranches are not all about trophy animals; there is a major component of animal management. A lot of the culling is done at a discount, and your customers are able to harvest their own protein. I know this may be a stretch for some, but it's another use of the land and a viable industry. In our area, there is an added benefit of such ranches. We have a growing population of feral hogs that causes enormous damage to crops every year; they are a non-native and invasive species. By opening up your property to harvesting these animals, you are also helping out your local farming community. For added value, the hunters might also like a class on how to process their own meat and other parts of the game such as the hide into leather.

Another way to define agritourism is by the season.

Some folks do an entertainment-style Easter egg hunt. Others might have seasonal workshops on plant identification including wildflowers and edible/medicinal plants. Seasonal events lead people to think about the holiday season.

In North America, Halloween is celebrated with pumpkins, and pumpkin patch time extends to Thanksgiving and into the Christmas season for tree sales. Many other niche holidays are built into this time of year. Some might give thought to other fruits and bounty to include observing a season of harvesting berries—whether it be blue, black, straw, cran, goose, or ... The list can go on for berries and other tasty produce.

With a commercial kitchen on-site, you would be able to add to your farm's offerings with cooking and food prep classes during the appropriate harvesting seasons. In that same kitchen, you can process your own produce like plums, making for value-added items to sell in your on-site store, or you can prepare meals for guests or bring in a local chef to do the cooking. With an on-site commercial kitchen, you can bring the farm and table much closer together.

Special events are another defining category.

Consider a farmers market and offer a spot for local producers to gather and provide a variety of items for visitors. Bonnie Chapa did this and included all the aspects of her operation. They brought in different vendors and coordinated shearers and veterinarians to tend the animals of local llama owners. This also gave non-llama owners a chance to see what it takes to own a camelid.

An open house allows you to draw people to the farm to see what you do. While your produce and products can bring people to your place, an open farm day can lead you in new directions. Since people want to stay longer, offering a countryside meal has been growing in popularity. People are willing to pay more because they feel connected to

the farm where the food was grown. One bonus to the farm-to-table movement is your patrons have the opportunity to take a tour of the garden before the meal. In developing a plan, determine how you can increase your visitors' connection to your farm.

Other ag-themed items that draw in people are just for fun such as a playhouse that looks like a combine or a seesaw made from old metal tractor seats. Carnival games fashioned after a farm's activities also help connect people.

Examples of special events can be found across the country from goat yoga to equine therapy. Some properties have added nature hiking trails to help people get out and back to nature. So many possibilities exist that, with a little research and commitment, you can find an agro-business you enjoy.

We delve deeper into one-time events in the next chapter.

Hospitality and short-term rentals are a subset of agritourism.

One rancher couldn't work the ranch anymore, so he remodeled his barns into short-term rentals. Staying on a farm is growing in popularity. Whether you're staying over for an event or you have an off-site obligation, staying at a farm sounds a lot better than a hotel.

During her farmers markets, Bonnie also provided an open house for her short-term rental property, The Silo House, so people could see firsthand all that's included when visitors stay there.

Off-Farm vs. On-Farm

Off-farm vs. on-farm experiences can offer some unique opportunities. Take your farm experience to an event or to an organization to showcase your operation in a farm-to-table encounter.

Many agritourism farms operate year-round; usually by stringing together a number of events to create an annual calendar. If it's part of your intentional design, holidays and agricultural harvests can span all year long.

Another spin on opening your farm is to consider who brings the programming. When you host a class on composting, you are providing the education. There are businesses that will provide their own curriculum and use your location to set the tone for the training.

We can separate education into its own section of agritourism.

School groups are a great place to start in the area of agriculture education as they concentrate the visitors into scheduled times. And if you can handle 30 to 100 elementary youngsters on your farm, you can handle anything.

Other types of tour groups are great for the same reason. You get a prepaid, predetermined number of attendees, with a set time limit for the visit. And remember, it doesn't have to be a tour of your entire farm. Maybe you have specialty animals, or you just really like ducks; in either case, you can teach people about your favorite animals. Maybe you have sheep or goats and it's birthing season; people love baby animals.

People want to know how to grow food more naturally, and, as Bonnie discovered, llama droppings can help them do that and you can teach them how. That's just one example.

If your operation has a way to include interactions with animals, it's a plus. Society as a whole is removed from their food, which in turn means people are disconnected from agriculture and production animals. This makes these animals a novelty to most. Animals belong on a farm, even if it's simply the ranch dog roaming the crowds or your pet potbelly pig checking out the guests. Animals bring character to your ambience. We might interact multiple times a day with our animals, but some people really do not know where their milk and eggs come from, much less their protein.

This also lends itself to the education component. A subcategory of education is the teaching of skills. The how-to lessons on composting, blacksmithing, improving soil fertility, raising quail, etc. Teaching classes about what you do takes it a step further with more details, which, in turn, gives a better experience. Teaching skills is also how people learn to do more with the produce or products you generate on the farm. People could learn how to make cheese from the milk you produce. Or people could gain skills in making elderberry syrups or elderflower teas from the plants you tend.

Seasonal harvests of any type provide year-round content and curriculum. If people knew more about how to eat locally and seasonally, I think we would be stronger as a whole.

Matt Wilkinson of Hard Cider Homestead offers people a chance to pick out their own turkey. Patrons learn how to raise a turkey, how to bring the turkey to maturity, and all the things it takes to get a turkey to the table. With guidance and help, people process their own bird for a holiday dinner.

People want to know more about where their food comes from and where it has been.

CHAPTER 5: ONE-HIT WONDERS

Can you make a living one event at a time?

It is possible to make a full-time income from doing one-time events. You may need to do several a year to keep the income flowing or open for a short, themed season. In our area, small towns hold annual festivals celebrating everything from rabbits to corn to fun, and more. Some of these festivals are centered around a harvest season. How can you bring a festival to your operation?

Single events like harvest times have been celebrated for thousands of years.

Why should your crops be any different? Towns and cities everywhere celebrate local harvests, from peaches and pecans to corn and cranberries. Most of these festivals include arts, crafts, and food vendors, and often have live music with an emphasis on the harvest of choice.

One of the considerations to hosting a single event is the time and energy it takes to produce, especially when Mother Nature decides to rain that entire weekend. So you have a single event planned for a holiday weekend and it's rained out, all the expenses still must be paid.

There is a benefit to running an operation for a particular season; you are able to extend your liabilities over several opportunities for income. As with anything, agricultural timing and Mother Nature play a major hand in what you are able to accomplish. For us, the bluebonnet season can last up to six weeks or as short as two weekends; we have seen them both. Another thing to consider with an outdoor wildflower venue is they do not bloom on command and scheduling your event for a blooming weekend is difficult.

Turn to Chapter 6 for more on seasonal events.

Single events take a lot of planning and can be challenging when it comes to making money. From a marketing standpoint, single events can generate awareness of your business, but more on that later.

Whether you are adding to or starting an agri-business, don't limit yourself.

Your location could become the destination for any number of events. Remain flexible to see these opportunities, whether you're putting on the event or charging rent for your space. Putting up a large pavilion, constructing an open-sided shelter, or keeping some of your field shredded for large tents may just make your farm *the* event destination. Even if you don't have a cool, old, refurbished building, some organizations or event planners will bring their own tent and supply all the equipment; they just need to have the space available.

James Halvorson, Halvorson's Hidden Harvest, told me a story about a rental venue that was an overnight peach orchard. It was a backdrop to hold corporate retreats with horseback riding. The owners of this location literally dug up 15-year-old peach trees, planted them at the new site, and had peaches the following year. One can only imagine the cost of having an instant peach orchard, but it became a success.

With an agri-business, you have rules and regulations about what you can do. Learn the dos and don'ts for your area.

Rent your property for special occasions.

Youth organizations are looking for places to hold events or allow the youngsters to work on badges and develop skills. Your place may be the perfect opportunity to help these groups. Granted, most youth clubs don't pay at the same scale as businesses and private parties but having a non-profit on your farm can be a great tax write-off and those are important to any business.

Weddings and, believe it or not, funerals are special events that might work for your farm. For funerals, if your location allows it, you might find a market for these occasions. Weddings are one of the single events that can help keep your operation going, but you'll need a steady flow of them.

When your business plan includes hosting weddings, keep up with what styles are trending. It may be popular to have a farm chic wedding; then again, depending on your locale, people may not be willing to be married in a barn. You must design your business carefully when you are getting started, especially doing specific one-time events.

Remember, weddings can be emotional for some people. The one coordinating the event needs to be able to deal with extreme emotions. On the flip side, you can simply rent the space for someone else to deal with getting people down the aisle.

Another thing about having a wedding on your farm: You do not get to pick and choose who gets married. If you have reservations or beliefs about people getting married, then keep them to yourself

or do not host weddings. Weddings can be a high-end event, and it takes working with people and plenty of personnel to be successful.

You may need to hire specific employees to handle event planning, especially for one-time events like weddings, because of the amount of follow-up and follow-through it takes. An extra salary will have to be calculated into the bottom line for these types of events. Employees are a critical point. If you are operating with minimal employees, one person out sick—much less severely sick—can, over time, bankrupt your business.

Music is another great reason to come together. Having a music festival that showcases local talent can be a great way to help budding artists get recognized and get your name out there as well.

Birthday parties are fun, and parties on the farm are the best. We've hosted all kinds of parties. Some were movie-themed, one was a costume party in October, and others were simply farm-themed because of the perfect setting. I've seen private parties held in conjunction with a public pumpkin patch. The party took place in a separated area, but the partygoers had access to the entire patch and received a free mini pumpkin.

People may just want to get out and enjoy the outdoors on a homestead. For these times, you might hold a special weekend and focus on your favorite animal. In our area, we have a landowner who hosts an annual Highland Games. A little farther away is another themed event touting the nation's largest Renaissance event. This one is counted among the overlapping events because they run for a short season and not just a single weekend. I felt it was worth mentioning here to help generate ideas.

Family reunions are good in the off-seasons. You are generally providing a place to meet, and the family will take care of most of the details. You just need to open the gate. Having a location between two major metropolitan areas can be a plus because families can meet in the middle. I've talked with people who plan their reunions around holiday seasons because they have time to travel, and they can pick up their annual pumpkin or Christmas tree and update the photos of the family.

Church retreats are like reunions, with the organization taking care of the details. The difference is when the organization wants to have a speaker or a team-building workshop. You'll need the infrastructure and specific equipment like microphones and audio speakers. With higher-end clientele, you might be asked to provide an audio-visual setup in a controlled environment. Corporate retreats are a selling point for new hires, and corporations are adding value in different ways to entice those new employees. As you build this side of your business, you need to be clear on what services you provide and then grow into what seems appropriate for your operation. Be flexible and keep an open mind.

Educational events and workshop weekends make for great one-hit wonders. A way to host single events is to remodel a building on your property into a place people want to be. With a bit of imagination and creativity, you can turn an old building with limited value into something of greater value, not to mention adding to the bottom line.

Matt Wilkinson of Hard Cider Homestead told me about how he was inspired to build a pizza oven at one of the Mother Earth News Fairs. He wanted to build an oven that other people could build for themselves. So, Matt went to work and then became proficient at making pizzas in his new oven. For Matt, making pizza and having a farm go hand-in-hand: His field-to-table experience includes picking groceries out of the field and creating a pizza to be cooked in the on-site outdoor kitchen. Being only feet away from where it's growing creates an intimate relationship with one's food. Matt's pizza-making experience is often used by corporations as a team-building exercise; employees are put in a social environment and work together to create a unique eating experience. Matt reiterated one must have the willingness to try new things.

NOTES

CHAPTER 6: A SEASON FOR A REASON

Why do people open their farms for a season?

Seasonal events could be related to holidays or times of year, such as harvest time, and they often overlap. Matt Wilkinson believes that I'm collecting the stories of these farmers because we all come from different angles of the same industry. Each of us specialize in our own markets of agritourism.

Seasonal holiday events come with a wide range of products, services, and, most importantly, experiences.

Your seasonal event will be based on what you're currently growing or working on and the goal for the occasion. If you are a pick-your-own, you are, of course, tied to the harvest. When your operation is educationally based, then you are tied to the time of year of your subject matter.

Specializing in one seasonal event, for sustainability, keep in mind how you plan to use your farm the rest of the year. Your focus might be corn mazes, but what are you doing the rest of the year? You need to keep the income flowing. So, it's a good idea to look at other opportunities throughout the year that might interest people. Depending on the size of your customer base at any given time, it may be best to aim for a few smaller gatherings with less income per opening as opposed to one big season for an entire year's worth of income. You may need to diversify for sustainability.

The last three months of the year include a major group of holidays.

Starting in October in North America, we have Halloween moving into Thanksgiving and Hanukkah into Christmas and Kwanzaa. Halloween prompts the need for pumpkins and photo opportunities with a variety of harvest arrangements. Many operations offer a pumpkin patch where folks can select the perfect gourd out of the growing field; however, some farms will need to bring in the orange beauties. Pumpkins might allow your site to provide the opportunity to carve jack-o'-lanterns with prizes for winning participants. And after visitors clean their hands, they can go on a haunted hayride. This time of year rolls into Thanksgiving with the fall theme continuing with pumpkins and adding turkeys.

The Christmas season brings evergreen trees and wreaths with bows. Some farms allow you to cut your own tree; others offer the chance to pick from previously cut trees. Leftover pieces of evergreen can be woven together to create a personalized wreath. The crafts that fill your gift-shop shelves might be themed with stars and nativity scenes. The experiences could include light trail rides and Santa's village.

With a little planning, you might be able to have one season lead into another.

Consider your region.

What are the local holidays? What celebrations happen locally that people like to attend? In a small town just west of my place, people gather to celebrate rabbits at the Rabbit Fest. In a town even smaller southeast of here, people gather to celebrate corn at the annual Corn Fest. All major and minor holidays and seasons offer an opportunity to reach people.

Easter is a wonderful time to draw people to your location. On our acreage, bluebonnets bloom around the Easter holiday. This spring tradition is perfect for capturing photographs among wildflowers or with a rabbit delivering eggs. No matter how the eggs or treats are delivered, children love scrambling to hunt and collect these springtime treasures.

Farm-camps can be viewed as seasonal.

Farm-camps can be for a single day or weeks long in the summertime. However, this also happens during the busiest time of year for farmers: summer harvests. Spring break can also be a time to host a camp on-site. Depending on your local school district's calendar, you could have one or two weeks of spring farm-camps. Matt Wilkinson produces softneck garlic and hosts garlic braiding classes, usually during spring break.

Field trips can happen any time of year.

Field trips, however, tend to follow a pattern: a few weeks in the spring and a few weeks in the fall. Be aware—when working with school groups—of the importance of having a curriculum associated with the farm and in line with the grade level of the students visiting. This gives you credibility with the district and will gain you favor over other operations. In dealing with school groups, be sure to have enough proper bathroom facilities for special-needs children. Making sure signs are clearly posted and all liability waivers are in place *before* the bus stops will help keep field trip challenges to a minimum.

Hopefully, the school will help facilitate the waivers; Don't let anything fall through the cracks. Another consideration on your site is areas for the buses to turn around and park.

Harvesting Animals vs. Plant Harvests

Why include animal harvesting in agritourism? Many lessons can be taught around the tasks of taking protein from the land and water, whether you show proper bovine butchering or guide someone to bringing in their first feral hog. For plants, we're including flowers, vegetables, fruit, and nuts.

Bringing home the bacon and other forms of protein happens in many ways.

With livestock, the animals are going to dictate to some degree when you butcher. Wildlife will have their own seasons regulated by a government agency.

It is an ethical must to be humane as possible when dispatching an animal. You need proper tools and skills to process an animal efficiently. Your tools and skills need to be sharp to make the end as quick as possible. If you're not familiar with any part of the process, find a mentor to guide you. In this aspect, books and videos are a good place to start, but having someone alongside you who has been through the procedure is even better.

There can be great emotional involvement when it comes time to harvest meat. For me, the kill is the worst part of my outdoor life. I take it extremely seriously. I have been through it in the woods, on the lake, and on the farm. It gets a little easier over time, but the feelings never truly go away. I know it is my responsibility to care for these animals from beginning to end.

The process of care extends all the way to the freezer, dehydrator, or whatever preservation method you use. When it comes time to harvest meat, take care to reduce spoilage. For larger animals, processing might be limited to winter months. If you have a walk-in cooler, you could process meat year-round. If you have a cooler the size of a classroom, you could teach butchering year-round as well.

Domestic animals have few restrictions on processing; the activity may have more to do with your ability to prepare the meat. Larger animals, like a cow, mean you will need ample freezer space or the ability to barter with your neighbors for other commodities. My grandfather, Pap, used to tell stories of the neighbors getting together for "hog killing time" in late fall because of the lack of refrigeration. A few chickens will not take up much room, but you need to calculate how many meals that will provide for your household. For Fran Tacy, owner of Franny's Farm, turkeys were huge for her operation because everybody wanted a fresh turkey for Thanksgiving.

Hunting wild animals has its own calendar with specific seasons and bag limits. Non-game, exotics, or non-native animals have different rules depending on species and location. Acquiring wild fish has similar regulations. Some species of fish kept on-site, whether in aquaculture or in your pond, will also have regulations imposed on them.

I remember Popa Kiefer, an outdoors mentor of my youth, always told me to wait until the first hard freeze to start taking cottontail rabbits for the table. In our area, the rabbits are likely to have a parasite that dies off after cold temperatures hit, but it's harmless to humans.

Some may consider wild animal harvests iffy when it comes to agritourism, but with short-term rental options and value-added experiences of teaching game processing, this method of harvesting protein fits right into my definition.

As part of an agritourism operation, plants can be split into many categories.

Flowers can be either wild or domestic; your location could be acres of tulips or other flowers waiting to be cut. Maybe your pasture lends itself to early flowering, but there may be more public interest in walking through areas of blooms during the later spring. This, too, can be in conjunction with a fair of sorts.

In Texas, we have bluebonnet season, and the state sees countless cars traveling the highways and byways specifically because of these wildflowers. What are the flowers in your region that you could add to or focus on in your agritourism business?

On my homestead, we celebrate the bluebonnet as a native wildflower by giving tours through the fields while discussing local ecology and farming practices. Of course, there are many opportunities for photographs.

These types of tours are a great time to instruct people on foraging spring plants for food or medicine, like Dewayne Hall on Red Feather Farms. Every season brings the anticipation of finding new plants. Maybe it's the time of year to hunt mushrooms; know which mushrooms come out first in your area. Then again, maybe you add growing mushrooms to your income stream and visitors pick their own mushrooms and learn more about the various types that grow in your area.

Nuts, vegetables, and fruits can be sold whole, as pick-your-own, or as value-added products. Allowing the customers to pick the produce themselves offers a unique experience. If you, the farmer, harvest the produce, consider hosting a farmers market on-site and include other local growers. Pick-your-owns are covered in another chapter in greater detail.

If you operate a pick-your-own or a greengrocer-related farm and have done your research on the regional varieties of produce, you can be more specific on the length of your seasons. You know the time to start harvesting strawberries and when strawberries will be finished.

Grapes have a widening market due to the possibility of a value-added product—wine. Even small vineyards host their own fair days; some regional vineyards team up to have wine trails, allowing people to travel to each location to tour and taste.

NOTES

CHAPTER 7: TIME TO PICK-YOUR-OWN

Having people harvest their own produce adds to the experience of being 'on the farm.'

We'll now dig into the topic of pick-your-own operations to give you a sense of the different types of such farms. Pick-your-own operations must fit the owner's business model and goals as well as maintain the infrastructure, integrity and profitability, all while dealing with the public. Another angle is community supported agriculture (CSA), which is food gathered and delivered straight from the farm to subscribers. Some CSAs allow their members to select their own food from the field.

Pick-your-owns can be successful; however, they need to be unique and well-designed to be a viable enterprise. Some pick-your-own operations like pecan orchards will buy back the extra if you collect more nuts than you need. Say you only need five pounds of pecans and you gather 20 pounds, you might get a discount for your extra labor. The orchard will then sell the additional 15 pounds to customers who don't want to reap their own bounty.

Before starting, know your property.

Matt Wilkinson of Hard Cider Homestead compares buying land to a first date; he says it takes time to get to know one another, and properties are as unique as individuals. You try something on a plot of land that isn't what the property likes, then you're going to need to work twice as hard to get the desired outcome. But … if you follow the patterns of the land and work in harmony with the land, you'll only work half as hard to get a better result.

When farming, you need to know prevailing wind directions and drainage profiles. It's not necessarily what *you want.* It is, however, what Mother Nature has provided for that property, and you must respect the land. The property will tell you what it's capable of but you have to listen.

We're all at the mercy of Mother Nature. Your blackberries had a bad year or your peaches didn't survive that late freeze? You might be in trouble. I've seen how that works. A few years ago, we didn't have enough cold weather, and the peach season went from a few months to two weeks. Your harvest times are definitely going to change, but with the proper design, you can hedge your bets.

Know your audience, know your industry, and know your area. To take it a step further, in a pick-your-own operation, it comes down to the microclimate. These microclimates can differ by more than a hundred chilling hours from one neighbor to another.

As James Halvorson of Halvorson's Hidden Harvest explains, most people don't know what chilling hours are for fruit trees — these trees need a specific number of hours of cooler temperatures before they will set fruit the following season. Buy varieties based on the recommended chill hours for your area; you want the widest range of chill hours to maintain the longest harvest possible. James urges you to understand the individual microclimate or microclimates of your property and the adjoining properties. You can use this knowledge to your advantage.

Pick-your-own operations charge for the experience.

When selling produce and you charge a flat fee for a particular size basket, it eliminates the need for commercial certified scales and inspections of those scales to sell by the pound. This makes for fewer headaches. We will look in detail at charging pros and cons in Chapter 10.

John Moody of Some Small Farm says people are willing to pay a limited amount for a product, but they'll pay more for a great experience. Adding to the interaction in the field gives your patrons something they cannot get anywhere else, and that makes your place stand out. For instance, allowing guests to sample freshly pressed apple juice from the fruit they are selecting will add to the emotion of being on-site. Customers will take away a true sense of your operation, and they'll talk about it and share it with everyone. Another example is a Christmas tree farm. Some may not consider these trees as a pick-your-own crop, but folks do it every year. Making it a family tradition, they'll bundle up, grab the Yule saw, and make a trek through the snow to select their perfect Christmas tree.

Offering true involvement can also be seen as adding value to a simple trip to your farm. These encounters will keep customers coming back again and again, and, next time, they're likely to bring their friends.

John says, "It's one thing to offer a product, it's another thing to offer an experience. You want them to walk away with a mark from your farm in the time they spent there." You want your farm story to become a part of their personal narrative.

Pick-your-own operations are tied to harvest times for your region.

Be very deliberate in determining your hours and specific times allowed for farm openings. People undoubtedly will show up at any time they want to; most simply do not understand that the produce they buy is truly seasonal. After all, those veggies are available in the grocery store year-round, aren't they? Education is key.

When your business gets big enough, think about hiring a dedicated office person to be the scheduler and oversee those appointments. In the case of a group showing up at 11:30 a.m. when they were booked from 11 to noon, to be respectful of your operation, your staff will need to hold them to their exact time and end the appointment at noon. Be upfront and communicate the reasons clearly.

If there seems to be a major complication with the shortened visit, consider allowing them to take photographs with the animals for a few minutes as a bonus. Anytime you can safely allow clients to interact with the animals and take photographs is good public relations; however, you need to have complete control over what is done around your animals. These creatures are in your charge and should be handled with care and dignity by everyone.

Pick-your-own operations require advertising.

Be very deliberate with your advertising. As I want to limit the number of patrons on my farm, I shorten my advertising window. It also helps me maintain control over how many employees I need. Over the years, I've been able to calculate the number of people to expect by the number of weeks I advertise.

Signs and directions must be noticeable and timely; include a detailed voicemail message on your telephone. Keep the website up-to-date and have a good Q&A section where onlookers can find the information they need without talking to a person, particularly a person who should be in the garden pulling weeds. Be active and deliberate in keeping everyone informed. It's especially true when you don't know the exact date when the strawberries will be coming in, etc.

Social media is an effective way to update supporters on things that happen spontaneously, particularly when you have good social interaction with your audience. You'll be able to let everyone know about your bumper crop of blackberries and that you've decided to do a special pick-your-own event.

It would be helpful for you to plan the entire year and set your schedule so people know well in advance. For some types of P-Y-Os, by-appointment-only works well; it's important that it's a time that is convenient for you. It's an even better idea for your visitors to pay upfront and know exactly when the allotted time ends.

Pick-your-own operations require education.

Allowing customers to select their own fruits and vegetables comes with the uncertainty of knowing if they will respect your plants, and they won't destroy your upcoming yields. Unattended children might, by accident, ruin a year's profits.

Consider requiring patrons to become certified pickers before they're allowed to harvest their own produce. Like any other endorsement, this will help educate what it takes to do something correctly.

Hold a 30-minute class on how to select whatever you grow. Teach beginners what is considered ripe and what needs to be left on the vine. In talking about this concept with Dewayne Hall of Red Feather Farms, we came up with the idea of a lanyard and badge to be given to verified pickers, and if a person didn't bring their credentials, they would have to repeat the class before being allowed to enter the field. It's an inconvenience to the customer but might save your crop. And some might see the accreditation as a badge of honor.

Your audience needs to understand regional harvest times. At grocery stores, most produce is available year-round, and everything must be shipped with a lot of fuel miles. Locally grown foods are limited to a particular growing season; some people don't understand this concept and will demand watermelons in winter. As you own a pick-your-own, set yourself up for success by not allowing a single group to come in and plunder the entire crop. Set parameters and make sure everyone knows what and when to select from a certain plant. Having earned a certification, patrons will know the harvest guidelines for your crops.

Think outside the box in designing a unique experience.

Bonnie Chapa of Laughing Llama Farm told me about an uncle who owned a property down the road about 25 years ago. He had one building with living quarters upstairs and a gift shop downstairs, and another held a restaurant. He'd stocked fish in several tanks and ponds, and maintained orchards with peaches, pecans, and other types of fruit and nut trees, so, naturally, he included a pick-your-own.

His operation was set up so a visitor could go fishing, clean the fish, and take the catch to the restaurant, where it would be cooked right then and there. Visitors could also eat whatever was on the menu.

The gift shop provided direct sales. People came to pick the peaches and take them home. An opportunity here was missed to teach canning and preserving techniques because they didn't know what else they could do with peaches.

Matt Wilkinson of Hard Cider Homestead mentions students picking their own ingredients as part of a class on mud-oven pizza making. Whether it's a pizza class or a tomato processing workshop, there are more ways to add to the experience. Outside of classes, for an added price, Matt W. allows his short-term rental guests to go into the garden and select what they would like for dinner, come back to this home-away-from-home, and prepare their own meal in the outdoor kitchen as part of the farm experience.

Anything you can grow can become a pick-your-own with the right design and planning. The biggest question is, "Does it fit your business model?"

NOTES

CHAPTER 8: LEAVING THE FARM

Can it really be agritourism if you leave the farm?

The business owners mentioned in this book come from varied backgrounds in the same industry. A couple said there was no difference between taking your produce to a farmers market and taking your farm on the road. The farmers market is simply selling goods. But most said if we take our farm's core elements with its spirit to the masses outside our fences, then that can be agritourism.

In my opinion, yes, you can take agritourism off-site; however, I need to set parameters around this statement. For me, it is still agritourism if your activity is off-site because you are carrying your farm's experience with you. Having your business model such that you can go mobile, all the better. The more options you can open up, the greater your chances for success.

For a number of owners, being on-site is necessary for it to be agritourism. Some get into agritourism to share their ag-operation with the public in addition to sharing the products and services generated from their physical location. Some people believe if the farmer leaves the farm, it can't be considered agritourism. In other words, it's your personal definition.

Dewayne Hall, Red Feather Farms, says all this depends on how picky you want to get. He teaches foraging off-site in an area called Post Oak to allow more options. Post Oak is a land conservancy with hiking trails through diverse mini-ecosystems that make great settings to teach regional foraging and plant identification, which is still in line with what Dewayne teaches on his farm 20 minutes away.

For example, when it comes to hiking, some trekkers are complete purists: Say they are hiking the Appalachian Trail, they must walk every mile even with the creek flooded. Then again, other folks are okay with taking an alternate route for safety. Sometimes you can go to other locales to teach a topic, but if it's completely dissimilar from what you are doing at your home base, then look closely to determine if it's part of your agritourism story or a different type of enterprise.

Showing your animals at a school carnival or setting up an educational petting zoo at a youth meeting can show people what your operation is doing while getting your name out in the area. By taking your animals to the area nursing home to let your animals do some public relations, you are adding value to the lives of individuals who may not be able to go to your farm. Or take the field trip to a school by bringing your animals on campus to include teaching small lessons on the care and maintenance when owning farm animals. Generating publicity in these ways can really kick off word-of-mouth advertising. Having an educational platform to bring your agri-experience to groups is a win-win. The community gains new experiences and your business benefits from the exposure.

The main idea is to figure out what works best for your business model. If it will allow you to succeed, then that's what needs to happen. Don't be narrow-minded, that will only limit you and your business. I know it sounds simple, but you want to do more of what brings in the money.

In Matt Wilkinson's case, at his Hard Cider Homestead in New Jersey, he feels if a visitor takes a portion of his farm when they leave, even just photographs, that is agritourism. One of the things he does at their short-term rentals is to recommend a bicycle trail of the area. This trail was coordinated with other area farms to be a reasonable bicycle route. The guests go off-property and experience other operations; they might stop at a nearby vineyard, learn about winemaking, and pick up a bottle to bring back to their temporary farm-home to accompany the pizzas they bake in Matt's outdoor oven. It varies on how creative you can be, and at the same time developing cooperation with the farms in your area.

As a national speaker, I strive to build community everywhere I go. I talk to everyone—from the attendees, vendors, other speakers, and cab drivers around town.

I expanded on an idea I had after listening to Matt W. speak. I have my audience members turn to one another to talk about why they are there and what they hope to gain from the presentation. This cultivates commonality and community. When we look back in history with respect to barn raisings or hog butchering, it's about individuals coming together.

For Matt and I, we can help our neighbors down the road be successful. Whether my guests learn from my neighbor or me, it's important people have those experiences to take with them. I'm including these interviews with other farmers to make our village of agritourism stronger. My friends, fellow speakers, and I have had our experiences, and I want to share those wins and losses with my audience to include *you*.

from Bonnie Chapa, Laughing Llama Farm – Troy, Texas

CHAPTER 9: COME ON IN AND STAY AWHILE

Having people stay on the farm is a great thing ... or is it?

A number of individuals will skip straight to this chapter to see if this book is worth the price. It's a growing trend to have people stay in your spare bedroom, guest house, or even a renovated silo. During my talks about agritourism, in this section, I start with a series of questions to get the audience to realize what they are asking for as they invite outsiders in to stay.

Do you mind having strangers on your farm?
Do you mind having strangers in your barn?
Do you mind having strangers in your house?
Do you mind having strangers in your kitchen?
Do you mind having strangers in your bathroom?
Do you mind having strangers use your toothbrush?

These questions might seem a bit extreme, but they give you an idea of your own comfort level. If your location with your personality can accommodate guests, then lodging visitors will probably be a good fit for you to open your farm to overnight stays. Ask yourself if you really want to rent out a short-term space. Next determine what the break-even is per night to know whether the venture would be financially acceptable.

Don't limit any option you may have when it comes to your land. If you can't bear the thought of an unfamiliar person in your house, then renovate the barn to allow the guests to stay on the property. Because you may not have a suitable outbuilding, consider moving one to your location. It might not be a typical structure; it could be something unique like a renovated grain silo as Bonnie Chapa did. She calls it The Silo House at Laughing Llama Farm.

New "apps" are being developed all the time for short-term rentals. A new one popped up on my phone the day before I wrote these lines. The trend appears to be narrowing the demographics to the sites that host "staying on the land" opportunities. This allows individuals to target the specific type of operation they want to go to, whether it is a llama farm, dude ranch, or a peach orchard.

As an aside, having a short-term rental, the masses will generalize and call it an Airbnb. To be clear, Airbnb is a brand, a business through which a person can book a short-term rental. For instance, everyone refers to a facial tissue as a Kleenex; Kleenex is the actual brand name of a company.

Many in society are not blessed to live in the country amongst the peace and quiet. A part of James Halvorson's family lives in the city, in high-rise apartments where they hear only police sirens singing with car alarms. The hustle and bustle of the big city echoes each day. So, when they come to visit Halvorson's Hidden Harvest, it takes them a few days to adjust, but then they start to say things like, "I feel like a pioneer living out here!" Having individuals and groups stay on-site is a wonderful opportunity for education when you have the ability and the personality for it.

Some people are willing to work for their room and board.

Dewayne Hall notes there is another way for a person to stay for a time; they could work for their stay. The labor option comes with its own set of criteria. Many variations exist for working on a farm to gain skills and knowledge. What could you teach an apprentice? Do you know welding, carpentry, animal husbandry, gardening, fruit-bearing plants, preserving food, or herbal tinctures? Because teaching is easier than using your back, teaching may appeal to you. Also, how long do you want students to stay alongside you as a pupil? You set the boundaries; you set the tone for what happens at your place of business. Such a visitor would need to offer a high skill set and be willing to learn other skills along the way.

Even with a short-term rental as a getaway, a few places allow visitors to sweat on-site as part of the overall encounter. This could sound like free labor, but realize you don't get to hand-pick guests. Prospects might say they will work hard, but they may not understand what the real grind entails. This leads to more of a one-on-one demonstration on your part rather than a hired hand to help lighten the workload.

Make farm experiences up-close-and-personal for visitors.

Particular travelers are simply looking for a site to set a tent or park their camper. This option is a fantastic way to start generating income as you expand the stay-on-farm leg of the business. Dewayne bought goats from a nearby rancher who had a small rock house he rents as a bed and breakfast down the road from his personal house. It's near where Dewayne comes up to feed the herd every evening. The visitors who stay in the little farmhouse love feeding time because they feel they're on a real homestead, plus that experience is the rolling definition of agritourism.

Fran Tacy loves having people on the farm, just as people love coming to Franny's Farm. She has her operation set up to build community. Everyone shares the kitchen, including being responsible for cleaning up after themselves. This self-reliance holds the short-term residents to a higher standard. Therefore, they communicate and learn from one another. Some folks are used to manual labor, but most are not. Certain patrons are active workhands while others are just there to vacation. Not everyone knows how to start a fire in the pit, but in a communal setting, there is usually someone with that specific skill who will pass along the knowledge. Amid this open atmosphere, you may have couples meet and later get married on the farm. Tenderfeet that first come to the rural life are, at first, dependent on others, then they eventually become independent to teach the new skill. That is truly what you want.

The best practices will likely take time to figure out for welcoming guests to your farmstead, but, ultimately, it's worth discovering things like self-check-in. If people are able to stay on your farm autonomously, not requiring you to do everything, all the better.

There is something powerful about connecting individuals to the land where they learn new things while interacting with others doing the same. A growing majority are truly disconnected from where their food comes from. However, with farming-based activities, we can bring the unaware back to an understanding of the stewardship of the land, including a map to their food.

Consider off-grid vs. on-grid for your short-term rentals.

As Americans, we have no idea of the conveniences we live with every day. Fran talked about her travels to other countries like Peru where they don't flush the toilet paper. In the United States, we use drinkable water to flush the toilet; that is our most valuable resource being flushed away. I hadn't thought about that concept until writing this book, but we have not been trained or educated on fundamental living. This may be an ideal educational platform on which to focus your own agritourism efforts: off-grid living or simply how to live better.

Fran started her short-term rentals off-grid with a robust solar station for visitors. She had posted signs on every outlet, "No hair dryers." Guests were provided French press coffeemakers that did not draw as much power from the mini electric grid as a conventional coffeepot. The heedless would ignore the signs and use a hair dryer, which would kill the entire power setup. Times became so extreme, they had to put in loud generators that used a great deal of fuel, both of which defeated the original off-grid plan. This grew to the point where they were ready to shut it down. Thankfully, solar power has come a long way since that time. Just be aware that even if people know what not to do, they may do it anyway simply because they don't believe the signs apply to them.

Your farm is a working location, and people need to be aware of that fact.

You may offer a guest house on your dairy farm, but it's a working dairy with a guest house, not the other way around. Your business may not include education on functioning dairies. This is where you need to set hard boundaries with strict guidelines. Bonnie Chapa suggests if you have trouble accommodating guests on your farm, you might need to take a closer look at your operation to ask yourself what is not crystal clear about the expectations.

At The Silo House, Bonnie provides a couple of fishing poles including a tackle box filled with lures. However, she is not a fishing guide. If her patrons want to go fishing, they are more than welcome to do so. But they need to have functional knowledge before they get there. Bonnie can offer advice, but she won't be able to teach them how to fish.

Bonnie loves to talk to everyone about her farm's main draw, the llamas, to pass along information about these long-necked lawn mowers. But feeding the animals is off-limits in most areas for obvious reasons, which is posted on signs everywhere. She has trained the llamas to not take any food beyond a certain fence. The llamas are smart enough that they know if there are humans in the pasture, they're not going to be fed. The sheep are even trained so two-legged field roamers will not get run over by food-loving animals.

People must be made aware that there is a designated area for feeding. You must be extremely clear with the rules for the safety of both your visitors and livestock.

Go into the rental business with your eyes open.

When you are willing to learn, grow, and enjoy sharing space with others, then short-term rentals may be a good fit for you. Bonnie says she absolutely loves it, and she's comfortable with this style of agritourism. But she did her homework, asking questions along the way.

Bonnie feels all challenges from the public can be managed if you look at your method of communication with your guests and make them feel comfortable. Bonnie recommends you tell them up front that your operation is a true working farm so there will be unique sounds and smells. Be upfront, be concise, be clear from the beginning.

Inside the parameters of short-term rentals, you will need to think creatively and expect the unexpected as there will be frustrations such as an 8-year-old unlocking the chicken house, letting loose 25 chickens. Oops!

Know that things happen, but with the right frame of mind, you will be able to find something positive in every situation. Each venture will come crawling with its own complications, but if you have a good outlook, the positives will outweigh the negatives. Attitude can be the key to success.

Most farmers have great outcomes with people staying on their property. I love talking with my colleagues to learn from their know-how. It really helps to have a growth mindset and be willing to develop and learn from someone else's trials.

NOTES

SECTION Three

THE BUSINESS OF AGRITOURISM

The numbers are where one finds success.

WAGON RIDE
PUMPKIN
U-PICK

CHAPTER 10: HOW DO YOU MAKE MONEY?

Here are the foundations of agritourism income.

One of the biggest challenges in business is to know how much to charge and make a profit. If you charge too little without taking into account all the factors, you may go bankrupt. If you charge too much, you may not sell enough to cover your costs. You don't want to take advantage of people, but you *do* want to know what they are willing to pay for what you offer.

You live on a farm and feeding the goats is a chore, but to someone else, it's an experience. They would pay well for their family to do your daily activities. How much is it worth to them?

Developing fees for your farm goes beyond material costs and labor. To complicate matters, there is no one-size-fits-all approach and some of the best strategies are multi-level.

Your message needs to convey your purpose and values. You first need to identify these core beliefs before you can communicate them effectively. You'll need to be honest with yourself when looking at the goals you've set for the business side of what you do. What do you want out of what you're doing? How do you want to sell your goods and services? Just because the neighbor does it this way, that doesn't mean it's right for you and doing it differently could set your operation apart.

What is your business model and its parameters?

Are you aiming for luxury, low end, or somewhere in between? To answer this, you must be clear on how you want to make money, your revenue target, and what you want to achieve. Goals and priorities will impact pricing, which in turn influences how your company is viewed by the public.

Be creative when setting your prices. If you find a gift shop price is too high, before you lower the price, raise the value by bundling items together. This raises the perceived value and moves inventory without losing profits. You're able to meet the customer's expectations and even over-deliver at the same time. These strategies will strengthen your bottom line, but they will happen only by design with an open mind.

One beautiful spring day, a young family was saying goodbye and thanking me for a wonderful day on the farm. Their little girl peered out from behind her mother's dress, then moved to latch onto my leg in a hug of gratitude. I was a bit shocked as I smiled down at her. She said, "Thank you for the best Easter egg hunt." This had me speechless. I had never fully grasped that my business could produce such joy in someone. This true story is a driving force behind my desire to see that every person with a farm includes agritourism in their operation. It's also why I say agritourism is worth more than money.

In traditional business, fellow owners are treated like enemies.

This isn't how I operate. Be true to your heart, and we will be stronger together. However, know your common companies because you still want to know who is in your direct market. In agritourism, it may not be another farm; it may be a business or event that vies for your customers' attention and time. It may be in a different industry altogether.

A movie theater can be in your direct market if it pulls people away from your event.

Keep an eye on new businesses and events that might draw away from your market. Watch old ones that change their pricing. When you can draw a correlation between what you do and what another operation does, you can use their pricing as a starting place, but it's not set in stone. Improve offers by combining items together to hit a different price point.

When you sell goods, your new customers will be looking at their options. What makes yours better? Do you stay stocked and ready to deliver? What value do you bring that they want or need? Know your customers through market research; the more you know, the better you will be able to provide what they desire or must have. Have your regular patrons bought into your farm story and wouldn't dream of shopping anywhere else?

Value is created in the minds of the customer.

Picture it: You're visiting your favorite pick-your-own; smell the sweet blossoms in the air. While you are filling a second large basket, a cheerful employee presents you with a glass of freshly prepared juice. Can you taste the sunlit nectar? Then they offer you a small cup of preserves made from the same fruit you are picking. They leave saying, "Thank you for visiting the farm."

Sure, the produce has a price, but involving visitors at different levels of your farm experience builds a relationship that turns a simple trip to the farm into a part of their story. Give the patrons something they cannot get anywhere else.

Increased value translates into increased profitability. I'm not suggesting you gouge anyone. Be reasonable and be fair. However, if you can't keep up with the customers, then a price increase will thin the demand. Having different levels of offerings will lead to diverse pricing that will meet customers at the different price points they're willing to pay. When a customer wants to spend more money with you, there needs to be a way to allow that to happen. Think of it like a silver, gold, or platinum level.

Purchasing psychology plays a huge role in customer decisions. People buy on emotion and justify with logic. Buying something means you filled an emotional need whether it was health, safety, security, or any number of emotional triggers. The action is then defended with logic, "I bought groceries this weekend, we won't be hungry this week." The truth is you feel secure in having enough food in the house to feed your family.

The same principle applies to buying a ticket to your event. "I bought a ticket, this better be worth the price." This is where you must over-deliver; provide goods or an experience above customers' expectations. Since they can buy a movie ticket and popcorn to be entertained for a couple of hours, your operation should at least do that much for the same price.

What are the benefits your customers gain by stopping by your place? What can you add to solidify—in the customer's mind—that your experience is worth the money spent?

Price is a function of your ability to sell. By getting them to feel, smell, visualize, and hear what it's like to be on an adventure with you, then they will not just believe but *know* your price will be worth it. In your marketing and advertising, what makes your operation stand out? Add one or more of the five senses to your description and mentally take them to your farm to make it come alive.

In focusing on value over price, look at the significance of your goods to the customer's expectations and then to their wants and desires. What meaning do your patrons place on your wares? Regularly examining both costs and values of your products and services can improve your profitability by maintaining the fair market pricing. At times, you will have to raise prices to remain viable and you might experience a drop in sales. This dip can be countered by being honest. Telling your story and the truth behind the increase can help buyers be more forgiving and more accepting of your need to bump up prices to stay open.

Remember to consider the value of what you are offering when setting your margins. We take our automobiles to the mechanic for what they know. They know what is wrong by listening to the engine or how to find the fastest way to fixing a problem after the computer pinpoints it. This specialized knowledge and tools are what we pay for, not necessarily the turning of a wrench. People pay to see farm animals in petting zoos, which has value.

Meeting the customer's expectations aims to provide what they are willing to buy. This may involve thinking beyond the status quo. To connect with the customer's possibilities, build different packages to fill their desires.

You can add an animal photo-op with a $50 purchase or provide a basket of produce if you bring a friend to an on-farm cooking class. Bonus items do not need to be expensive to add value.

Value-based pricing takes into consideration intangible emotions and is a dynamic process that can ebb and flow with the times. If you can add value to a basic item, you should, but know your true costs to be effective. Cost-plus pricing is the simplest form of figuring fees and takes the least amount of thought. You take the actual cost of an item and add a percent. You can be successful with this style of pricing, but you must know your exact costs.

Keeping an eye on costs is a necessary part of business.

While there are different types of costs, they all fall into two categories: fixed or variable. Fixed costs are accrued regardless of productivity. Variable costs increase with output. The same expense can fit into either category depending on your business model. By adding fixed or overhead costs plus the variable costs, you arrive at the real cost of an item. Once you understand this true cost, you can decide on your profit margins.

Total Fixed Cost + Total Variable Cost =
Total Cost

Volume is another term you will hear associated with cost analysis. Production volume is the number of something you make. Sales volume is how much you sell. If you sell Christmas tree ornaments, production and sales volume will not be the same due to inventory. However, when you sell a service, you produce and sell at the same time, so the numbers match. (Example: hours of operation.)

I made 100 ornaments = Production Volume

I sold 75 ornaments = Sales Volume

I have 25 ornaments = Inventory Volume

Our Christmas tree farm will be open for nine hours today. = Service Volume

Fixed costs will be there whether you open the gates or not, like mortgage, rent, utilities, leases, loans, salaries (not hourly), insurance, and annual taxes (as in property taxes). These expenses are based on a set time period:

30 days use of a building = rent

30 days use of money = loan

2 weeks of time = salary

Salaries are paid in spite of number of hours worked or amount of manufactured goods. Hourly wages are usually based on varying amounts of time worked, leading to profit creation, therefore associated with variable costs. Using this definition compared to your style of business will tell you if the cost is fixed or variable. Some items may crossover to be variable if your business uses the item for production.

Take precautions against adding fixed costs to your business. Fixed costs do not end with the season, making lean times difficult. Raising fixed costs increases the need for steady monthly income in order to pay the bills.

Variable costs increase with production, services rendered, or the number of hours the farm is open. The hours of operation is a specific period of time; it's your hours of service and can vary.

In making peach preserves, your cost increases per jar. You'll need more jars, lids, labels, raw materials (peaches), and hours to prepare the finished product, but your potential income increases as well. Expenses rise with the variable cost as does the capacity for profits.

What is the market value of your product or service? Like the mechanic, it's what is offered to the customer and what they're willing to pay. Don't hike prices for the sake of profiting, meet the clientele at their comfort zone of value. I would rather people visit the farm multiple times a season then only be able to afford one trip.

Remember: Cost is the money it takes to produce X, the price is what is paid for X, and the value is what the customer is willing to pay for X. Price should be in line with your customer's value.

Costs increase when the business has to pay more for merchandise or services. One example is a hayride—is it entertainment or is it an educational farm tour? When consulting with a tax authority in my area, we discussed the various activities on my farm and the conversation turned to hayrides. This seems simple enough; put people on a trailer and haul them around while they sit on hay bales. He then asked me exactly what I did while we were out on the trail. I told him we slowly ride along until we get to a spot where there's something to discuss; we stop and talk about ecology and what's happening on the farm at that time of year. He asked, "So you're educating the public?" I answered positively with all the enthusiasm of a recovering schoolteacher. He said it comes down to definitions. If I'm only hauling people sitting on hay, that's entertainment and taxable, but if I'm educating the public, the ride would be non-taxable. The key word is *education.* Check your locale for similar advantages.

Side note: In our area, straw is cheaper than hay, so we use straw bales covered with light saddle blankets. This has led to zero allergic reactions compared to using baled grass.

Make sure you know what is earning a profit by carefully analyzing each of your offerings. Simply stated, do more of what is truly working and adding to your bottom line.

Once you have a sure handle on your total costs and know your break-even point, you need to home in on the market value to see if you will be profitable. You'll need to know your market well to know what the public will pay. Gathering this information and other statistics is called market research.

Market research is as simple as knowing your customers.

Paying attention to their need—and, more importantly, understanding their wants—is key to getting into the minds of your ideal clientele. There are a great many ways of collecting this information. As a start, simply listen to them when they speak. Take it a step further and ask them questions. The ways to ask vary greatly: on-site conversations and quick suggestion cards to follow-up email blasts for signing up to your blog.

These techniques also apply to potential customers. In your advertising and marketing, draw out responses from people so you know who is seeing your message. Listening and responding shows you care, and people want to be heard. This type of feedback can be invaluable in guiding your business to success. Given a chance, consumers will tell you what they want to have more of.

Feedback and market research should not be a one-time thing, otherwise your information will become stale and dated, which in turn could lead to bad decisions. Get in touch with your customers often and let them know you are in tune with what they are saying. Public recognition for giving feedback in the form of coupons or discounts gets people's attention and helps others speak up.

Market research can include other businesses in your industry, in addition to your own customer base. Knowing the local trade can give you a starting point for building your operation and pricing structure. Shopping your specific market will allow you to see others' designs and create a guidepost to what's going on. Fair warning: Just because someone else is doing X or has a certain price point may not make it viable. They may be trying something out or simply not know something is losing money.

I did a test for a friend; the idea was to do an Easter event because no one in the area was running one at that time. I have always maintained a smaller operation, on purpose. On our farm, we do more than agritourism and want to keep the gatherings family-sized. My friends don't mind 10,000 patrons in a weekend. They encouraged me and did some of the marketing to jump-start this concept. I knew I was a guinea pig for their operation to test our local market. We were in the same geographical location, similar types of operation, just different demographics. This was not a financial threat to either of our businesses, and what's more, we were able to work together.

Knowing your demographics and sticking to them will help you hold your course. We all may want to be in an upper-tier customer base, but your location may only support mid-level pricing. Having an ideal customer in mind when you do or change anything will help you maintain your direction. Change can be good but changes need to be tested, especially when it comes to setting a price.

What is your time worth?

For some, they give it away. For other people, $250 per hour is their base fee. This is only part of the art and science of pricing. Finding the midpoint between your wants and the customer's wallet can enhance how much you sell or how you operate.

Hobbies cost money and businesses make profits. The bottom line should be to make $1 over cost. Turning more than that leads to greater sustainability. John Moody's opinion is that the way to make money in agritourism is to create a product or experience with a value greater than the cost of what it takes to produce it. James Halvorson says knowing your absolute bottom line is a way of knowing what you need to charge to break-even on any enterprise. Then find out how much your customers are willing to pay for each item or experience.

The simplest pricing would be to slap a dollar amount on what you are doing or making and hope it works. Next would be to know your actual costs and adding a percentage on top of that to see if you can stay in business. Ultimately, you want to know what your market will pay for your value and charge accordingly.

In planning, develop a budget while keeping a revenue target in mind. The budget is a list of incomes and expenses of your business into the future. The target revenue is what you want to financially make. An action plan is how you're going to get there.

Income – Expenses + Target Revenue =
A Positive Number

(If it does not equal a positive number then you must adjust one of the three factors.)

One way of charging guests is to have, as part of an admittance fee, a few low-maintenance activities like playgrounds or shelled corn pits, and then up-sell for things that require extra labor or for the cost of crafting materials.

Part of the action plan is to test prices: new offerings, new combinations of longstanding products, sale prices, and marketing. When running sales, do not run them too close together, or the sales price becomes the norm and you'll have trouble getting your margins up, especially with your long-term clientele. This also can be said about changing prices. Changing prices too quickly will trigger your steady customers to question the upsurge and maybe shop around, when they normally would not. There is a flux in sales when you change a price.

Your supplies might have gone up and you don't want to lose money. You raise the price $4 per jar to cover the cost of new raw materials and pad against any future increases. In this case, expect a drop in sales. The increase dips the number of jars, but the higher price covers the loss. Change prices incrementally only with good reason.

A best practice to price increases is to be honest. If materials went up or you now offer more to do on your farm that requires more employees, then say so. At the cash register when people balk at the new price, inform them of the reasons and most will understand given that the increase is rational. To support your decision to raise prices, you might give the added benefits, list the cost increases for your raw materials, or supply chain issues. Then again, some folks will never be happy, and they are expensive to keep around, so your business may be better off without them. There are ways to lessen the discomfort of spending more money out of pocket.

As lowering prices in general is not best, we have focused on price increases. Price reduction should be temporarily managed with "Sale Pricing" or seasonal price markdown to move inventory. Low prices suggest low quality or cheap.

How do you take the sting out of inflation or pricing methods?

One approach that we all have seen is using odd values such as $9.99 as opposed to $10. This works effectively, but you should weigh out the perception. A combination of both might look like $10.99 to enter the fairground and home-canned pickles are $11 per quart. The entrance fee is minimized by the 99 cents, but the pickles are premium as a whole dollar amount.

Charging a flat-dollar amount for a basket of produce also works to qualify your offering as superior and avoids the need to weigh it out on a scale, which can simplify your accounting. Having different sizes of baskets available also gives the customer options depending on their individual needs.

A loss leader is selling something at cost or less to bring people in and making up the deficit on other items. This can also be accomplished by giving something away like free admission for youngsters under 11 when you know they are going to eat, drink, and ride every ride. This has a perceived value to the paying adults. You could also run a buy-one-get-one when you have a bumper crop or a free item with free paid adult tickets.

Wholesale pricing works as long as your operation can maintain the volume required to allow people to buy in bulk. A build up to such pricing is a tiered system where the more you buy the more you save.

If you absolutely must raise prices abruptly, consider adding a bonus item with the new price. The new package item could be something of less value to the business (like a small basket of a perishable crop that might be wasted if not moved quickly) but the value is seen by the customer.

Run your business, don't let it run you.

In agriculture, factors like weather and calving season often take precedence over other plans. What you can control is what you offer your customers. You set the hours of operation and the calendar of events. If you operate a farm store, you don't have to be open 8 to 5. When you're the one behind the cash register, make the times convenient for you. Remember, you own the farm. (Only opening from 6 to 8 p.m. Tuesdays and Thursdays will condense your time out of the field and still get your merchandise to customers.)

In a loose sense, I think of my bank account as a scorecard. If I total enough points, then I can open the doors another day. When I don't tally enough, then I need to revamp what I'm doing.

Looking deep into your business activities to ask where you can add benefits for your customers will lead to higher profits and greater success. That sharp pencil will keep pricing accurate between the business's needs and the buyer's wallet.

It is critical that the value is felt by the purchaser. Retain your pencil's sharp point with meticulous monitoring of markets, costs, and values. You owe it to yourself, your business, and your patrons to keep prices at the front of your business mind.

NOTES

NOTES

CHAPTER 11: HOW DO YOU MAKE MORE MONEY?

Value adds extra income to your operation.

When I was a public educator, one motto was to "beg, borrow, or steal." Of course, this was said tongue in cheek. I wouldn't recommend it as a business practice. We all have great and unique theories on how we can better utilize our farms. It's part of my philosophy that if we learn of a particular concept from someone, we need to take that design and make it our own. Not to steal somebody's idea, but to take it in and turn it into something uniquely yours. Don't poach, transform. I learned this lesson by jumping into a pumpkin patch blindly. I thought because they could do it, so could I. I was wrong.

Two peach orchards can thrive next to one another if they stay true to their own passions and collaborate rather than compete. It's a way to help make each operation stronger. One orchard might make peach wine. The second owner may love to sell jellies, jams, and preserves along with offering a pick-your-own operation. It's a perfect example of how two farmers with the same product can serve different markets while existing along the same fence line.

As opposed to a single source of revenue, value-added products from your farm or locally sourced farms offer numerous revenue streams. Moving your operation to the next level, you take your farm's raw foods and turn them into something of greater value like cheese, yogurt, jams, jellies, salsa, or wine. Don't forget the full array of non-edibles: soaps, lotions, lip balms, etc. An on-site store is a fantastic way to show your offerings, but due to government regulations, you'll need to follow local rules and guidelines to get the goods to your customers.

Focusing on the experience people have at your farm is crucial to success.

Imagine walking into any store where you normally shop. One day, a person just inside the door greets you with a smile and asks if she or he can help you find anything. This changes the dynamic of your shopping experience. The same thing is true for your operation. Add value by adding to the experience. If you are not the most outgoing individual, hire an extrovert to welcome people to the farm. If you have multiple-day events or a particular tourist draw in your area, staying on-site might be next-level for your patrons.

Your operation, by design, should have various levels of connections. For a country birthday party, the first level might simply be the location. Level two includes a farm tour. Level three might increase the experience by adding pony rides. Every step includes more activities and increases income. The same progression could be applied to weddings and other events to increase people's interactions and your profits. Evaluate your goals and business plan to be sure you're on course and staying true to your vision. This will keep your operation growing and your clientele coming back for more.

When you sell a cucumber, you make a little money. When you sell a pickle, you make a little more. When you sell relish, you make even more. Then you hold classes about growing cucumbers, canning pickles, and making relish, for greater profit still. In this way, you're able to sell your cucumbers and teach the classes to provide the ultimate experience; people get to take a piece of your farm home and duplicate the product. Patrons are able to repeat the process at home and are then able to tell others about your farm.

Sometimes you can make more money by not spending any money.

Our hay crop is a by-product of managing our wildflowers for our spring events. We cut the hay at the right time of year to maximize our yield of bluebonnets for the following spring. We don't spray anything on our fields; no herbicides or pesticides, no 'cides at all. By avoiding those amendments, we make an all-natural hay crop and are able to net more per bale by having fewer capital inputs.

By adding a commercial kitchen to your venture, you can personally produce mercantile products, rent the space for other farmers to do the same during your off-season, and teach more in-depth classes. This increases the valuation of your outputs and makes better use of your downtime. With this setup in place, offer dinners on the farm by providing the raw ingredients and partner with a local chef for a unique experience.

In finding people and organizations to work with, there doesn't have to be a direct financial component. Inviting a scout group to sell their cookies or candles at your farmers market could be a win for both. The group makes sales, and you are seen as a philanthropic farm. Always make sure working relationships benefit both parties.

In the same way we don't pour money into our hay crop, you don't have to sell eggs with an expensive label if your clientele knows your chickens. When people come to your farm and see your chickens free ranging, you don't need a label to tell them what they see. With a little education, people learn more about your flock and know the chickens get top-quality feed to supplement their free-range lifestyle. Marketing can lead you to get the same price or higher per dozen as a carton of "labeled" eggs.

When you have animals, it may be a promising idea to offer a service to others that you already perform for your own animals, such as hoof trimming or shearing. This leads to networking with similar-minded individuals and builds a community around your farm. Having specialized equipment is a write-off and can supply a service to your area. It might not tie directly to the agritourism operation, but it can definitely increase the profitability and sustainability of your total agri-dream.

Ask yourself two questions if you're looking to add something new.

Dewayne Hall knows how to crunch numbers, and he recommends doing your research on any new product or produce like elderberries. He recommends answering these two questions:

Do you have a market for your product?

Do you have the labor to get the product to the market?

If you are counting yourself as part of that labor force, remember how you fit into the equation with your age and time availability. You need to be devoting your time to developing best practices. When you're running the whole show, you need to oversee everything and not necessarily be pulling weeds all the time. Then there is your plan: What stage are you in your business plan? Do you want to be pulling those weeds or teaching a class about your favorite crop? In your homework, you need to discover how many products can be derived from the main produce growing on your farm. For elderberries, you can sell fresh berries, dried berries, flowers, and starter plants, not to mention any number of consumables. This one product can be enhanced by growing herbs to complement the making of teas, tonics, and/or tinctures, and by offering an almost endless list of classes on each. A bonus to teaching classes is that, when the students connect with the teacher, they tend to buy their supplies from the instructor.

It's a permaculture principle to stack functions in everything we do. If you place a water catchment system uphill to water your elderberries, develop a class on rainwater catchment systems. You can create a class on everything you do around the farm literally from underground up.

Your operation will determine added-value items.

John Moody of Some Small Farm says there are myriad ways to add value, you just have to find what works for you. If you have awesome wildflowers, add a photographer to take family photos. If the patch is large enough, cut a maze in the flowers. It's all about finding what people want and giving it to them.

James Halvorson says on his farm it's all about value-added products. His version of the pickle example takes him from cucumbers to a jar of pickle spears. If he took those cucumbers to the local farmers market, they'd sell for about $2. By cutting the cucumbers into spears, placing them in a quart jar with pickling spices, and processing, they sell for five and a half times as much. When you take that philosophy and apply it across your farm, you increase the value of your entire farm.

James offers an upsell by allowing people to come to the farm to pick berries, fruit, or vegetables, depending on the time of year, and offering a class on what they can do with what they just picked. Making strawberry jam, peach preserves, or watermelon pickles, canning tomatoes, making tablecloths out of old feed sacks, weaving grapevine cuttings into baskets and floor mats, or even using a sawmill are a few examples of a how-to class people might enjoy.

People are looking for an experience, which is an upsell. A consistent theme to being sustainable is to upsell from eggs to bread to a pet's overnight stay. (A word of caution: If you allow people to bring their pets to your farm, make sure everything is designed accordingly to keep your animals, your patrons, *and* all pets safe.) Check on the regulations of your state and county concerning interactions between people, livestock, and pets. Your area may require certifications, such as animal welfare and handler certificates. Enforce strict rules on where pets can and cannot be in any given situation. If a customer is going to bring their animal to your farm, there should be a charge to help cover any extra paperwork and certificates.

For general lodging, offer your guests a simple farm breakfast and coffee, but anything that increases your workload needs to be an added cost. When you

stay at Franny's Farm, Fran Tacy will let you know what add-ons are available: foraging in the garden, asking someone to go to the store for you, having a chef cook dinner, or experiencing an on-farm massage. This goes back to the collaboration with your local community, and it creates word-of-mouth advertising. The community advertises for you, and you advertise for the people providing those services in the community.

A farmer has a wide range of options when it comes to income streams.

Once people are on your farm and asking questions, educate them on the benefits of livestock manure; Bonnie Chapa found that llama droppings make excellent fertilizer. You'll make money from people staying on your farm but sending them home with a sack or two of high-quality fertilizer can generate a nice side income.

Bonnie uses feedback from her customers as a way of learning how to add more value to her farm. It's a way, she says, to find areas where you can grow and meet the needs of the public. Complaints may offer the best lessons; if people are happy, there is little to learn. But unhappy people will tell you what's wrong or what could be better. When nobody is grumbling, you don't know if there's anything that needs to be fixed or changed.

As another way to look at value options in your short-term rentals, are you adding fresh baked goods, offering a bottle of wine, or including fresh farm eggs in the refrigerator? Know what your time is worth, and how much the baked goods cost as opposed to the wine or the eggs and ask yourself if it truly adds to the farm experience. Take the time to do a cost analysis of these extra items. The eggs will be the least costly in terms of time and materials as a way of adding value. If, however, you're running a vineyard, the example will change, and a bottle of wine may be more appropriate. Look at the other farms in the area … but stay true to your vision and take these ideas to make a unique offer of your own.

On Hard Cider Homestead, Matt Wilkinson plants garlic, which produces a lot of scapes, the top part of a garlic plant. If you don't know what to do with that extra greenery, that's a lot of useless material. So, he uses those scapes instead of basil to make pesto. Making the pesto led to making pesto lamb sausage, which turned into a new product line with a personal recipe.

How do you take one product and find multiple ways to sell it? When a person starts thinking with a creative mind and a sharp pencil, that's when you start increasing income to your operation. With a short-stay rental, you have a fixed income.

80% weekend occupancy x 50 weekends per year = 40 weekends booked

40 weekends x $100 = $4,000 fixed income

So, take the opportunity to sell more products from your farm, which increases your bottom line and your sustainability. Find ways to diversify your income.

Matt W. says one adds value by adding to a person's initial interest in your farm. For instance, someone spends the night and asks you about a certain type of tomato. You say it's a good slicing tomato and, if combined with basil and homemade mozzarella, we can make tomato-mozzarella-basil sandwiches. The neighbor makes homemade buffalo mozzarella and the farm next to him grows and sells wheat flour. This is a prime example of an upsell, and you didn't give the customer more than they wanted, you just let them know what's available, and it helps you and your fellow farmers in the process.

Your role is to help your customers figure out what they don't know. You're there to assist them in learning what's available and how to put together all the pieces to make a full package.

With their rental, let them know that they can build a campfire, have eggs to take home, or that there's a number of farm products available. Look for things you can offer that, coupled with their original attraction, will enhance their experience on your farm, which will help you add more profit.

A lot of people don't like to do sales, but it's different if you're simply letting your customers know what's available in conjunction with their initial interest. Inform your guests, through signage, if you are unwilling to have that conversation, but your operation will be more profitable if the information comes directly from the farmer.

NOTES

Fresh
Brown
Eggs

CHAPTER 12: MARKETING AGRITOURISM

If people don't know about you, how can they come see you?

If you skipped to this section, you know the importance of quality marketing for any business. Check out the strategies in Chapter 14 regarding recent economic times.

Most people use the terms *marketing* and *advertising* interchangeably. However, advertising is a part of marketing, and marketing is much more than just advertising. Advertising is to inform people: "Farm Day this Saturday! Come on out and enjoy the fresh air." Marketing is finding the right people who will actually come to the farm and putting your advertising in front of them. As scary as that sounds, it can be done simply with a little knowledge and market research.

Take a look at radio, television, and print outlets.

Traditional media sources such as local newspapers, regional magazines, billboards, radio, and television can play a role in drawing attention to your farm; each has its specific niche. Traditional advertising can be expensive compared to other forms of marketing, and it may be a little harder to track the cash generation.

Traditional ads have a place in an agritourism marketing campaign, but with your venue in mind, are these efforts going to pay off with higher upfront cost? Bonnie Chapa warns of advertising with some traditional outlets because you may shell out money with little to no benefits. Be careful of auto-renew policies, a semi-annual or annual bill may suddenly appear and add an unexpected cost at a bad time.

Never lose track of where your money is going and don't advertise unless you can track the results. By tracking the gains, you'll be able to tell if you're putting in a dollar and getting two dollars back. You might discover that you're putting a dollar in and not getting that dollar back or worse: losing money altogether.

One less expensive avenue to get the word out about your farm is to do something unique that will make the news—like helping a local charity raise money or donating your space for an annual youth or civic meeting. Because you're able to do something kind, it will drive public awareness to your farm for almost no cost, and it's from a third party.

When you get a third-party recommendation, it's like word-of-mouth advertising, which can be extremely powerful. Think about it: If your friend recommends a restaurant, aren't you more likely to try that eatery? Third-party recommendations work the same way. Similar is true for online reviews, but we'll talk about those later.

Having a popular radio station broadcast from your location during an open farm day is one way to bring people out on the spur of the moment, but it's hard to quantify its usefulness. And if you put up a billboard, how are you going to measure its effectiveness? All traditional media can be used successfully if you can discover what your target audience was told and where the information came from.

Connect with your target audience face-to-face.

Interacting in person, on the farm or off, can be a major part of your marketing efforts, and it can generate good results.

A lot of people don't think about it, but business cards are a great tool if you always have them with you. How many times have you wanted to remember a particular place of business and didn't have something to write on or you snapped a photograph of the business' sign, and the photo was buried in your albums?

While it sounds dated, a business card can also be a conversation starter. Your operation should be always at the front of your mind; you never know when someone is going to need to know what you do. A well-designed business card can be a focal point. An extra tip for using business cards effectively: Have some preprinted with a discount or leave a space where you can personally write down a buy-one-get-one-free offer.

A good complement to a well-designed business card is branding, adding your farm logo to these cards, as well as to the shirts and hats you wear every day. It's important that when people see you, they also see your farm and how proud you are of what you're doing.

Offering on- or off-farm in-person demonstrations to school or civic groups can be a great tool. These presentations are more entertainment to pique the interest of the audience. If you have a concern about public speaking or teaching groups, it's an opportunity to partner with someone. Matching strengths with other farms should complement both operations and make each farm better. James Halvorson at Halvorson's Hidden Harvest likes the concept of marrying a product of the farm to an experience, then bringing in another expert educator to explain the whole process. Build a community by bringing in those with the skill sets.

One of the ideas that came out of brainstorming with James was a drive-in movie night at the farm. Because you have the appropriate site and set-up, this could attract attention to your farm, especially with seasonal movies centered around a holiday. Try horror movies in October, or Christmas movies in December.

Another way to draw attention is to give people what they want. Even if you have a small following, ask them what they'd like to know more about when it comes to your farm. If the consensus is something unfamiliar, then find a willing expert. By asking the people who already know your work, you tap into your market and gain more accurate information on what to add to your operation.

Acquiring new customers is one of the more difficult aspects of business, and paying attention to the people who are already there can increase your profits, quickly. If they've come out to your place to take a class on beginning basket weaving, the next logical step would be to offer a second class on advanced basket weaving.

A warm market audience is a group of people who know your farm, are actively following you, or coming to your events. The more time or money they have spent with you, the hotter they are considered.

When you have multiple events on or off-site, always let people know what's next. With a warm audience at one event, tell them about another; they are more likely to attend it as well. Keep reminders everywhere. It may feel pushy to you, but you never know what is going to grab their interest. When you have a minor event, create a display for your major event, and at your major event, post all the dates for your minor events. View each as a marketing tool for the next one.

This allows you to use a smaller event as a loss leader or a way to get people who come through the door to spend more on other things. The loss leader in grocery stores would be something like milk, eggs, or bread that the store is selling at or near cost. People go to the store to buy such staple items and then while there, they'll pick up other things. The store profits on all the other items. You can write-off the cost of a minor event to promote your big seasonal event. Your minor events throughout the year will keep people coming back and you stay fresh in their minds for next year when your big event rolls around again.

Take advantage of networking opportunities offline.

Networking is about building relationships, and it needs to be beneficial for both parties. Offline networking is more about personal contacts rather than mass appeal, though both are important to success.

One effective way to network offline is to set up a booth at local fairs. Know the demographics of your target audience and the event you're attending. If you're going to an event for health-conscious individuals, setting up a booth with your own healthful elderberry syrups or gummies would likely be a good fit. If I showed up to the same event with photo albums, showing the bluebonnets and families having a great time on the farm, it would still fit the bill for a health-conscious crowd if it was a family-oriented gathering. Matching up types of demographics will make your booth more cost and time effective.

From a business-to-business standpoint, a photographers' convention taking place in your area offers a chance to spotlight your farm as a venue for taking family photographs among the wildflowers. When selecting events to attend, match your products and services to the needs and wants of the event goers.

Setting up shop at a farmers market is a way to make sales while experiencing face-to-face interactions with potential visitors and collaborators alike. The other vendors at the same event will have similar demographics for their business and working together can benefit both of you. Don't stay behind your table at these events, mingle and you might find a win-win arrangement for your farm.

When starting a relationship, whether it's business-to-business (B2B) or business-to-customer (B2C), your goal is to create value for the other party. What's the perceived benefit of doing business with you? What are you giving them in advance so they might spend more time and money with your farm?

When it comes to B2C relationships, offering a discount code makes an upfront positive gesture. Another way to pay it forward is to give a referral discount to your existing clients. As they refer new customers to your short-term rental, you give them a price break the next time they rent.

Allow other people to feel as if they, too, are stewards of your land; they'll feel empowered to act as an ambassador, talking about their experiences on your property. A small group of people could even

Free
Worm
Tea

be trained to give tours of your farm while you are tending other aspects of your operation. You as the owner need to be doing the higher-level business operations as your farm expands. It won't be a one-person operation for long, therefore, you will not be able to meet every customer every time. A goal would be to have other people doing that for you, people who absolutely love your land. If you're doing all the leg work, then who's doing the brain work of building your business?

For over a decade, Matt Wilkinson at Hard Cider Homestead has not done a whole lot of off-farm advertising. He simply takes excellent care of people. Being sincere and listening to people are keys to great customer relations and meeting their expectations. Aim for top quality in your offerings, and if you make a mistake, correct it. The honesty in interactions with people and the willingness to make good on any mistake are going to keep people coming back, which leads to people talking positively about you and your operation. Strive to be a five-out-of-five-star operation and the rest will take care of itself.

One thing to keep in mind as your profits go up is that, generally, your variable expenses go up as well, and it takes a sharp pencil to keep those two in balance. When it comes to marketing and advertising, spending money should turn a profit. If the money you spend is not making you more, then take a closer look. Tracking every dollar is critical in any agriculture venture, especially agritourism. As you build your business, advertising and marketing become easier.

Networking online will likely reach more people.

Now for the cow in the room that some people love and others hate. Marketing digitally can be huge, but it doesn't have to be overwhelming. With a little understanding and taking it in small chunks, it can literally "save the family farm."

Networking is a type of marketing but online, it's more of a one-to-many opportunity rather than a one-to-one experience. For the most part, it's a one-way communication as it allows for a large number of people to gain access to you and your farm. As your audience builds, you'll be able to find prospective people to work with regarding projects that can be beneficial to both sides.

From a B2B or owner-to-owner standpoint, it can be crucial to your success to use your network or the people with whom you've built a rapport. Such a network may not be mandatory to succeed, but it definitely makes business more enjoyable, even if your network is spread across the country. You never know when one of your counterparts a few states away will have family and friends in your immediate area. Somebody from your audience may be traveling and because they're familiar with you, they are more likely to stop by a farm run by someone you've mentioned. This happened to me when I had a speaking engagement on the West Coast. Through social media, I contacted several farms along my travel route and visited them to gain insight on how things are done in different areas. From a business perspective, I could write-off a trip to visit someone else's farm to see something new and visit friends I've made in our industry. I'm open to learning more.

Just because someone's audience is hundreds of miles away doesn't mean customers won't find you through different channels. That's why there needs to be a balance in giving to other owners in the same industry. I don't see other people doing similar things as the competition but as complements.

Don't forget digital marketing and social media.

When I speak of digital marketing, the lines get blurred between true marketing to find your ideal customer and advertising to let them know what's going on at the farm. The online world can find a model consumer and provide advertising about things they want to know more about.

Social media has evolved into a platform with many sides. You can stay connected to family members across the globe or across the street. Here we're going to focus on the business aspects of social media, even though people telling their neighbors about the farm is what we want.

Digital marketing consists of digital ads, websites, blogs, emails, social media (Facebook, Instagram, Twitter, Pinterest), videos (YouTube, Vimeo), and podcasts (digital talk radio). The true genius of this digital world is how all these things can work together for you.

Marketing agritourism is similar to any other small business, and it can be expensive and difficult, but it doesn't have to be that way. These two facets can be someone's greatest weakness; however, if you're passionate about your business and honest about what you're doing, many avenues are available to start simply to build up your farm's digital presence. Applied correctly, one of the beautiful things about digital marketing is that it can be very cost-effective. One downside is that you have to be constantly creating some kind of content, so consider outsourcing your content creation if you are not up to the task.

Digital marketing in today's world is probably the best way to get your farm's name out among the public.

Nowadays, you can't have a conversation or use the word *marketing* without bringing up the concept of social media, which is a huge chunk of current advertising and marketing efforts.

From a B2B standpoint, you're staying connected and working with your network of people. If you help other businesses promote one of their classes or events, they should, in turn, help get your message out when you have a class or event. Otherwise, it's not a win-win, and it becomes a win-lose for someone.

This is not to say that because two farms are in agritourism, the other owner will be willing to help just because you ask. Some owners do not have a growth mindset and view everyone as competition.

The first step is to be of value or to create value for others with your presence, then turn that into a mutually beneficial relationship. To find these possible business partners, be active in your digital spaces. Constantly put yourself out there in a manner consistent with your personality.

To stay connected to your audience and patrons, you'll want to have a system that includes a website and email list. This then leads to you taking control of your own marketing efforts. Social media can expand this system, but the core effort needs to be simple: Get email contact information. By collecting email addresses, it becomes much easier to send out a newsletter or a single announcement directly to individuals who have shown interest in your products or services. It's important to stay relevant to your warm market and let them know what's going on plus what they can look forward to at the farm. With their permission, Bonnie Chapa collects information from her guests in order to communicate with them, and she lets previous visitors know what's going on at the farm.

Eventually, your own social media will be key, but there are different levels of social media marketing.

When it comes to posting on social media, you never know the feedback you're going to receive. This is where a thick skin helps; view every comment as constructive criticism. A complaint is an opportunity to better serve your clientele.

Your marketing should be frequently catching the attention of potential customers, but occasionally, you make an exceptional friend of the farm, a person who takes special interest in your operation. If you happen to make a friend who is particularly good with social media, and you can add value to their journey, pick their brain when appropriate to become better at your own social platforms.

With current technologies, the process of finding the right people to hear your message is becoming easier every day. Simply telling your story is one of the strongest ways to market your operation. It creates a connection with individuals who are more likely to spend money with you. Finding these people is what market research and marketing are all about.

Honesty is key. If you're not transparent, people will know and tell others, quickly. This is a two-edged shovel that cuts both directions. If you're not genuine, they will smell the cowpie. If you are authentic, then people will become life-long patrons.

Technology has changed the way businesses market to consumers. This shift allows owners to manage their own advertising campaigns without the need for an ad agency or public relations firm. But, if after a little exploration, you still don't like the idea of running your own marketing, through the same technology, you can hire someone to handle those efforts for you.

Part of my marketing motto is to "track the traffic." If you spend one dollar in marketing, you need to know how much income that single dollar is bringing into your business. Tracing your money can be done with all the traditional methods, but you'll need to filter the information through some type of system to monitor where the traffic originated. We must know what works and what doesn't.

Many email marketing companies are out there that offer free accounts and assistance with setting up your own email marketing campaigns—complete with templates and extras that will help your campaign stand out. Some provide templates for newsletters and flyers for specific events as well as customizing features to import your own photographs, graphics, or logo.

A benefit of using social media and email is that all these accounts can be consolidated into one location, which allows you to pick and choose the outlet and the section of your list to receive your newest campaign. This is going deeper into types of marketing but is not necessary in the beginning. The rate of response may vary according to the outlet you select. The slight differences between each platform is a million-dollar industry and you don't start there, you start with a website and an email list.

No matter how user-friendly the Internet has become, the biggest complaint is the time it takes to learn and work on all the pieces to this digital puzzle. One solution is to outsource your digital presence. This may be one of your tech-savvy children, hiring a part-time staffer, leveraging a friend of the farm, adding someone who literally works for eggs, or contracting a virtual assistant or VA. A VA is someone who works remotely for a set number of hours on specific tasks.

Social media is great, but it's not always the best for bringing in bigger deals. To close large transactions, you need to form a relationship where there is mutual benefit.

How will people find your farm's website on the Internet?

No matter what sector of our industry you are in, you will need to consider search engine optimization (SEO) for your website. SEO is making your website more recognizable for the various search engines people may use. Ranking on the Internet can be extremely important to your operation. If you're on a national short-term rental platform, you will have to earn a ranking and that takes time.

Bonnie Chapa gave this example: The last time you stayed at a large chain hotel; how did you find that location? What did you do to locate that room? Answering those questions will give you some insight into what you should do for your own short-term rental.

In all your communications—from website to email to social posts—be clear and concise about what you offer and your expectations on your farm. What will they get for their money? Matching what you offer with the customer's expectations is critical for a good experience and for them to leave a good review.

You have no control over what people are going to say, but word of mouth is still the most powerful form of marketing.

The rule of geometric increase says that if you get 10 people to stay on your farm and those 10 people talk to 10 of their friends, you just advertised your farm to a hundred new people at no cost. When those 10 people write a review, especially a positive review, then those 10 people turn into thousands.

NOTES

NOTES

Fran Tacy, Franny's Farm – Leicester, NC

CHAPTER 13: WHAT IS THE BUSINESS OF AGRITOURISM?

This is the boring stuff that will make your life easier.

The sad truth is most people won't get this far in this book, much less launch an agritourism business or expand their current operation to its fullest potential. Since you have made it this far, "Congratulations!" You are on your way! I would like to encourage you personally, so drop me a note saying you made it to Lucky Chapter 13.

We started talking about insurance agents and lawyers in the first chapter. Here's where we flesh out that discussion with more detail.

Disclaimer: I am not a lawyer nor an insurance agent. This is not legal advice but stories about what I've been through or what my agritourism friends have experienced.

Agents and lawyers have their areas of expertise, and these specialties can narrow their ability to take care of your needs. Even if a particular insurance agent or lawyer has been with you for several years, you might need to find someone to fit your current enterprise more effectively. You need to have someone who understands your market, and this understanding is imperative for your success.

If an insurance agent doesn't know how to write a policy to cover your agritourism/farm operation, you could literally lose the farm. If your lawyer can't specifically advise you on your business activities, you could lose the farm.

Insurance agents represent larger companies, which have their own definitions and guidelines to protect their own assets, not yours. You may have to interview a dozen insurance agents before you find one who comprehends your operation and represents a company that can take care of your needs.

It all starts with a name.

The following are things to consider when naming your farm. Some people use search engine optimization (SEO) techniques, while others use local history, and yet other people name their farms for personal reasons.

Consider the necessary legal structure when it comes to adding an operating name—"doing business as" (DBA)—as opposed to your legal name; this helps separate your various entities.

Dewayne Hall followed a more spiritual path when it came to naming his operation Red Feather Farms. He studied as an apprentice medicine man for more than 20 years and was ordained into the Native American Church. In his region of Oklahoma, it's an accepted practice to follow Native American themes, and this works well at Red Feather Farms where he follows more earth-centered practices.

One way to name your farm is to connect it to the theme of your operation.

Bonnie's Bluebonnet Farm

John's Pumpkin Patch

Fran's Apple Acres

When you read the farm's name, you know exactly what to expect. John Moody at Some Small Farm says the beauty in a simple farm name is that people will know what you're offering.

James Halvorson, Halvorson's Hidden Harvest, says it's your personal preference when it comes to a trade name. His farm name goes back to northern Wisconsin where his family had a piece of property tucked away out of town that they called their "Hidden Hideaway." It was a hunting and fishing camp where the family vacationed. James enjoys wordplay and the three Hs are easy for people to remember. A name that's easy to remember is always a bonus in your marketing and advertising.

When James bought a spread in Central Texas, it had a lot of the same characteristics as his original place in Wisconsin. His farm today sits at the end of a one-lane dirt road in the middle of nowhere surrounded by nature. In acquiring a new property, James studied the natural harvests from the land. Once you take a deep look at a property and really listen to the land, Mother Nature will let you know what grows best there. James also feels one needs a personal connection to your brand name, and it needs to be easy to say and remember because it'll be there for a long time.

Another way to identify a property is to dig into local history and learn about the land's past. The area's road and street names may be tied to individuals who had influence in the area or owned much of the land.

For Fran Tacy at Franny's Farm, it's personal since she's the farmer, visionary, and spokesperson. This is not the case for everyone because some people want to own the business but not be the public face. Fran recommends picking something cute, fun, or even silly to make people smile or elicit an emotion, but she says it also needs to be personal. Be proud of the name you hang on the gate because you're building a legacy and that is how people are going to remember your farm.

At first, Bonnie Chapa's farm was called Heart Rock Ranch because on the day they closed on the property, they found a heart-shaped rock in the driveway. Bonnie admits the first thing people want to do when they buy land is name it. As her operation continued, Bonnie renamed the tract Laughing Llama Farm because everyone who came out loved the llamas and people left laughing.

In my opinion, you need to be a part of the land before you give it a permanent identity. One of my favorite quotes is Shakespeare's "to thine own self be true." For me, this is important when choosing your agritourism theme and mantra, and the first place you must be true to yourself is in your farm name.

Matt Wilkinson of Hard Cider Homestead suggests people ask a few questions of themselves when deciding the name of their operation.

How do you want to be known?
What do you want people to think when they see your farm name?
Who are you as a person?
What are you trying to achieve?
Why are you doing what you're doing?
Why are you thinking about implementing something else?

He suggests being authentic because being insincere doesn't work. People will see through false efforts. Be honest and wholeheartedly believe in what you are doing, otherwise, you're just pretending. Genuinely analyzing yourself and your operation is a lot of work, but it can make other decisions easier.

The name of Matt's farm (Hard Cider Homestead) is based on the street name—Cider Mill Road—and Matt's unique take on life. His farm's logo is a jug with three Xs, the symbol of moonshine, and he says he's not afraid to blaze new trails and follow his own direction. The farm name, to Matt and his family, captures who they are and what they are about.

Carefully select a business structure for your operation.

Ultimately, you might have several ventures under one banner, such as a working cattle ranch and a pick-your-own strawberry patch. Because of the diversity, you could have each operation separated for tax purposes, liability, or other business advantages. These are our personal experiences and not legal or tax advice; *always* do your research for your specific situation.

Ask a hundred people about legal structures and you'll receive almost as many different answers. The contrasts are mainly due to personal backgrounds as people's history directs them when it comes to different company designs and life in general. You'll need to do some intense homework on what will work best for your operation.

Even when you talk to a tax adviser and a lawyer about how to structure your business, each will give different answers based on their individual past and education. My lawyer talked about liability and my tax adviser talked about covering assets; they see the world from different angles and neither one was wrong. You're the one who knows your business the best and in the greatest detail, so you're going to be the one to do the due diligence to figure out what legal structure will best serve you and your operation.

We strongly suggest finding professionals in each area of your business. The professionals you choose should have a working knowledge of your specific operation and follow the most current best practices.

I have some agritourism counterparts who swear by S corporations, and they almost hate LLCs (limited liability company) because of the low entry point to start and see the limited protection as false security. According to the Internal Revenue Service: "S corporations are corporations that elect to pass corporate income, losses, deductions, and credits through to their shareholders for federal tax purposes." A "limited liability company" should be reflected on its owner's federal tax return. The distinctions can be confusing to decipher but critical to get accurate for the best possible outcome.

Study the regulations (and fine print) for each business structure you are considering. For instance, pay close attention to what happens if the business fails. How are your debts going to be paid? Carefully read all paperwork that must be signed and forms that need to be filled out.

Each arrangement is going to have different benefits and drawbacks when it comes to taxation—Social Security, income, employee, sales, federal, state, county, city—and are operation specific. These are key points in determining which one will help you make the most money for your dream.

For example, S corporations have special rules

in filing Social Security for individuals and family members. This can be a money saver, and the way the profits are returned to these individuals can also help you retain money.

Some agritourism farms operate with a sole proprietor. Simply as another level of organization, add a DBA or "doing business as" certificate. The DBA allows you to open a bank account under the farm's name. Even though everything is under your personal surname, it makes it easier to track expenses and income from the business side.

James Halvorson filed a DBA through the county to merely protect his business name so it couldn't be used by anyone else. The filing protects the name for 10 years.

If you start as a sole proprietor and your enterprise takes off, you can, for lack of a better term, upgrade to a different structure. It can be harder to change structures if you start with one of the more advanced arrangements.

Look at funding when deciding on a business structure. Some grants are only available to certain demographics and/or business types. If you plan to seek one of these grants, it may be appropriate to select a more favorable business structure, which may include changing who is listed first as the business owner of your farm.

For others of my associates, a C corporation, which taxes the owners separately from the business entity, means big business with big dollars associated; that is something you're going to have to weigh against the size of your operation.

One of the simplest measures of what structure you should pick is deciding how much risk you are willing to take.

Though no one wants to talk about it, remember that businesses fail. Ask the hard questions of yourself and of your team of professionals, "What if my business fails?" By asking tough questions, you'll go through scenarios with each business type and determine what it will offer you in the way of protection of what you are building.

Again, I stress seeking the advice of professionals because they should know the details in your business arena. In my travels and discussions, I've found people who believed their neighbor or cousin and made decisions based on hearsay and speculation. I've heard many stories about what people think a business structure will protect and when hard times hit, they find out that it isn't true. Consequently, they lose everything.

Bonnie Chapa cautions of possible scams by people who are not licensed professionals; they'll take your money and deliver little of value when you're setting up your business.

Many of these suggested steps can be taken on your own for free or at a low cost. You save money by doing it on your own, but you'll need to know what your time is worth compared to letting an expert set your business framework. We go to professionals such as mechanics, dentists, plumbers, financial planners, and educators all the time for their special knowledge, skills, or tools for a job to be completed quickly and correctly the first time. These tasks can be DIY, but it takes time to learn the expertise these people use every day.

As part of your business development, take into consideration whether you will have employees, contract labor, both, or neither. Once you choose to have employees, you'll need to contact your state's Secretary of State to apply for an EIN or employer identification number. It's likely the process can be handled through the state's website, but the procedure will vary from state to state. Do your homework and see what makes sense to you: A sole proprietor with a DBA, LLC, S Corporation, or C Corporation with multiple business structures underneath each. These are just a few of the options. The final decision is yours. Take a look at what you really need and start there to build the business as you go.

When you bring people onto your property, assess how much risk you are willing to take and how you are going to minimize those liabilities with your business structure. Another important question to ask is, "Are you living on the property?" Your business filings should protect your personal assets as well as your business holdings. Be sure your company guards your home should someone sue your farm business.

Having release forms for classes, whether it's processing animals or cooking, is a best practice for any business structure.

To reiterate, you should find a mentor to help you walk through these steps, but you are going to have to make time to determine the correct protection for your enterprise and personal life.

Getting paid is an important consideration and needs to be settled early.

When I asked Dewayne Hall how he got paid, he said, "You think we make money at this!" He sells eggs for $3 for either cash or check. When Dewayne works other people's goats, it's either cash or check. He tries to stay away from credit cards due to the fees, and he was in the financial industry long enough to

from Bonnie Chapa, Laughing Llama Farm – Troy, Texas

see how credit cards have hurt a lot of people. From a business perspective, if someone buys an item for $300 using a credit card, then I'm going to receive less than if they would have paid with cash or by check. One of the workarounds for taking credit cards is adding a convenience fee. It must be termed correctly for your city, county, and state to be compliant.

Customers love using credit and debit cards but accepting cards can be a headache. You can add a fee to cover credit card processing or build it into your pricing.

James Halvorson recommends taking any form of payment even if it means setting up a merchant's account to process credit cards. He says you never want to turn away any type of payment. For the first two years James was in business on the farm, he only accepted cash and found that the people who went up the road to an ATM never came back. At the farmers market, people would simply go to the next booth because that vendor took plastic. He said a new business will need to figure out what is feasible and plausible for payment options.

He also recommends bartering for certain items after establishing the value for both sides, then making a deal where both parties are satisfied.

You can also charge by the package or by the pound. When you sell by weight, you will need a certified scale; don't forget to figure in the expense entailed to have the scale re-certified. That re-certification is one reason many people choose to sell by the package.

Fran Tacy of Franny's Farm suggests getting the money upfront as this obligates people to fulfill their end of the deal. People have become accustomed to just canceling reservations, especially when it comes to short-term rentals. Because of this, you need a very clear, distinct, and strict cancellation policy. Fran sees the benefit of using an established short-term stay website in the handling of back-office items. When someone pays, the money goes directly into your account. Often there will be a cost for this service with established platforms, but it streamlines many things. Other payment processors like PayPal and Square allow you to send and track invoices, helping to keep your business organized.

Fran says patrons at their pick-your-own operation pay for a basket upfront then fill it. If they need another container, they return to the beginning and pay again. She also suggests taking payments however you can.

The size and style of your operation or type of event may change how you get paid. Bonnie remembers doing farm events where the vendors set up for free. This allowed each vendor to track their own transactions with their own credit card processing and merchant accounts. She says you never want to be the money man in the middle when it comes to card processing.

One way to charge vendors is to have them pay 10% to 25% to you as the host although figuring out that percentage can become a headache. An uncomplicated solution to this dilemma is to charge a flat fee. However, even this so-called simple method can be tricky if you're just starting out and don't have data from past events. Vendors will want to know how many people came through in previous years before committing to you.

In the beginning, it may be worth it to hold a farm day event that is free to the vendors and/or free to the public, just to collect data for future paid events. Be sure to have someone dedicated to taking accurate counts of total attendees, total number of cars, and the like. Getting feedback from vendors and guests is important. Ask each vendor what they thought the free space was worth. That way you'll have a better idea of what to charge in the future.

Free-to-the-public events give you similar data on how many people are likely to show up. Don't get stuck in a certain mind frame where you cannot give anything away. You're weighing what is given against the data you can gather to earn money in the future. When starting a new venture, feedback and data are critical to making solid decisions moving forward.

Matt Wilkinson points out that when offering classes or demonstrations, a lot of people like to attend with another person for a sense of companionship and support, especially when the topic is new, and to make memories. In this case, giving a discount for two or more people may be a powerful incentive. Almost any discount can be recouped with strong enough upsells throughout the event.

Don't forget the taxes.

Every city, county, and state has different tax codes, and this is one of the most important areas to research for your location. For me, when I buy a fence post for the farm, it's just a cost of doing business and nontaxable. But for your operation, that same fence post might be taxable. The same is true for technology used to help advertise my business. It may be an expense but is it taxable to you?

Pay your taxes and know the possible deductions. Dewayne Hall uses a common computer program to help stay organized. Many software programs can help determine what is taxable and what is not, which is handy when filing. There is a computer-based program for any size business, even if you are running 100% cash.

Discover what you can take as a deduction on your taxes. Dewayne gave an example of buying a supporting gimbal to help do interviews and take video footage. For him, it was a write-off or a tax deduction. He also owns a bandana company that prints hiking trails, animal prints, and all manner of information on handkerchiefs. For him, going on a hike can be a write-off.

A qualified tax professional will have an eye on your books and will let you know where you stand with your annual deductions. Depending on your structure and your income ratio, your tax professional may tell you to spend $4,000 by the end of the year or you are going to have to pay $800 in taxes. This is one of those times where it pays to precisely know your finances; $800 is easier to come up with than $4,000. However, if you're in need of $4,000 worth of equipment, and you have the income to do it, then that might be an appropriate time for you to make the purchase. Finding a capable accountant is worth the cost and will save you money in the long run.

Sales tax is another area of the money game, and it's going to be very location specific. Your individual state will have a website and an office with all the information you need along with the definitions of what is taxable and what is not.

The IRS has a website with all the codes; however, I find it much easier to pay a professional to handle these situations.

Because I hire a competent professional, it saves me a week's worth of work, and I know I'll pay less than a week's worth of my time. For me, it's about perspective. A mechanic knows his hourly rate, a plumber has specific fees, even doctors and dentists know what they make. All these professional people know what their time is worth. You should too.

To help your tax professional, be diligent with your records. Keep a paper trail for everything you purchase by keeping all receipts. The easier we make the work for our tax person, the less it costs out of pocket. By dumping all annual receipts into a shoebox and put it on the tax man's desk, expect to pay more for his time. One tip is to buy everything using a bank card. That way you have a double paper trail.

Part of the tax burden is dependent on your business structure, and you may have different entities for each aspect of your overall undertaking. As an example, you have a bed and breakfast with a peach orchard. Each might need to be independent of the other for tax purposes. An individual enterprise may need its own unique tax certificate depending on the subset of agritourism you are operating.

When you get into short-term rentals or the hospitality side of agritourism, there may be specific taxes associated with that venture along with additional county, city, or state hotel taxes. When dealing with taxes, Bonnie Chapa suggests speaking to a tax adviser because you need to know what to expect. Do your homework on your specific commercial enterprise and seek specialized advice.

Matt Wilkinson points out that one of the easiest ways to avoid mistakes is to find a mentor and create a relationship with someone experienced in your category of agritourism, which can really set you on the right path.

The nature of your agritourism operation will dictate the sort of insurance you'll need.

These are guidelines I try to live by, and I hope they help you ask the right questions of the different specialists for the most complete coverage for your operation. It's a starting point to find the coverage you need.

Insurance is a risk management tool. If you had $50,000 in debt and had a million dollars in savings, there wouldn't be a need for insurance. If you had $50,000 in debt and nothing in savings, then you definitely need coverage to protect your assets if something goes wrong.

In the search for your ideal insurance agent, tell them the truth about what you are doing and what you plan to do in the coming years. As your business grows, your insurance needs will change to meet the new ventures. Farming has always been a dangerous occupation, and different parts of your business will have varying degrees of liability and risks when you invite the public into this lifestyle. In these discussions, make sure your coverage is specific to your individual enterprise and not just a general policy that may not cover what you need it to. A generic policy may take care of 80% of your needs, but you require coverage on the other 20% as well to be 100% protected.

A trained and savvy agent is a must, particularly in the case of any special event you have planned. You can contact your agent and have an umbrella policy drawn up to cover you for a certain time period. They should be able to give you a price quote for that event so it can be budgeted in advance.

Insurance, accounting, and legal matters—you need to have qualified professionals working with you to protect what you're doing.

Farmers need both health and life insurance. Know your options because going without health or life insurance could put your farm at risk. When factoring in the cost of insurance, do not overlook your eyes and teeth.

Self-funding your insurance needs means you have enough money set aside to cover medical bills or loss of life. However, I had a friend who was self-funded; when she was diagnosed with stage 4 cancer, she was prepared to pay cash for the treatments. The hospital she wanted to receive care from would not take her case, as she was self-funded. She finally found a hospital, but I found it strange that a business would not take cash.

Most farmers or spouses work off the farm as a way of providing health insurance. If you fit the demographic, you may qualify for government subsidized programs like Medicare or Medicaid.

You may have some options to receive health care through the Veterans Administration if you are a prior member of the military or a military spouse.

Health sharing or cost sharing works like a co-op for the members of a group. Each member pays a monthly amount that is distributed across the participants. Whenever you start calculating the cost of getting health insurance on your own or what your employer is paying for your health insurance, you might get sticker shock.

Health insurance may be the largest budget item. When developing a business plan, your health insurance must be considered for the long haul; with every passing year, the cost will increase.

Insuring the farm and the real property can be handled by a national organization or by a regional company that may be more familiar with your local needs. As our culture gets further away from the agrarian life, fewer companies offer insurance to cover agricultural needs. The limited options make it more important to shop for coverage tailored to your specific situation.

Certain companies, and more specifically individual agents, might not understand that the equipment in the barn is worth more than the belongings in your house. So you need to deal with people who understand farming and an agricultural way of life.

Another type of insurance you need to know about is product liability. Simply put, this is protection against damages caused by something you made or sold. If you work with government agencies, you will need to have a specific minimum dollar amount of coverage.

In some locations, there is an insurance policy specific to farmers markets to cover things like a vendor's tent blowing into somebody's car.

Liability is a big concern in business, especially in agritourism. When you have a working farm with people and livestock near one another, you better have good insurance in place. Some people see a farm and think the owner has a lot of money, and this mentality opens the door for lawsuits.

Some liability can be managed with proper labeling. A good place to start when preparing edible goods is under your area's cottage food laws. However, product labels are no replacement for an active policy.

The Texas Agritourism Act protects (Texas) landowners with limited liability. Most states have an agritourism act of some type, and to receive the limit of liability, you must post signage or have a signed written release with the correct language. It's a best practice to have the correct signage and signed releases along with adequate liability insurance when opening to the public. Check with your state government to know the specific requirements.

I have signage on the driver's side of the public entrance to my farm that lets it be known that this is a working farm and you enter at your own risk. It's specific and the wording is from my state's legislature. One sign applies to agritourism on working farms and a second sign focuses on horses and horse handlers. Proper signage is no guarantee that you won't be sued; however, having your state government back you is a good start. Other states have similar requirements where lettering and the signposts must be a certain size and in specific locations.

Waivers and signage can ease the burden of risk when you don't have insurance or when you can't find an organization to cover your event or activity.

Having an attorney familiar with farming—and specifically agritourism—is key to effective protection. Subscription-based legal services can assist you in drafting waivers and proper signage to protect what you own. When dealing with people, anything can happen, and you need to be prepared. If you are sued, win or lose, you will have to spend time away from the farm and this will cost you hours and dollars; either way, you'll be out a lot of both.

What are your insurance needs for structures, equipment, and features around your property? There is no one-size-fits-all solution when it comes to insurance; look at the overall scope of your setup and the underwriters may dictate how you operate. Nobody wants to be told what to do on their own property, but if you can't get coverage, you might need to change what or how you open to the public.

Hopefully, your agritourism operation is growing and expanding. You may not have all the attractions you want, but make sure to get the coverage for what you have right now. Visit with your insurance agent as your business develops.

You may choose to not have property insurance for your buildings, especially if you have the materials and are capable of rebuilding what might be lost. Assuming you're handy enough to reconstruct, you still need to weigh the insurance premium against rebuilding what you had.

In case your operation is small or livestock isn't your sole income, you may not need to concern yourself with insurance covering loss of animals. The same would be true for crop insurance. There is food safety insurance for processing food items. You need to know the details of your state's cottage food laws and the like while you're making products to sell. Most cottage food laws are geared to protect modest ventures. Labeling requirements can alleviate some of the insurance needs for these operations as long as you sell the product appropriately and your labels follow the requirements.

On the hospitality side, there are specific short-term rental insurance policies. As this is an evolving industry, stay current by independently researching and staying in contact with your insurance professional.

You may not be able to find insurance for everything you do on your farm. As an example, your insurance company may frown on an animal processing class. Moving forward with a class like that with no insurance in place makes it imperative that people sign a waiver and a release, documents preferably drawn up by a well-informed lawyer. In general, but particularly when teaching a novice how to use sharp objects, you will have to be extremely safety conscious when operating outside the boundaries of an insurance safety net. This is also when you analyze the risks and rewards of offering such a class. If all profits are going to insurance costs, then you must ask, "Is it worth it?"

NOTES

CHAPTER 14: SAVING THE FARM IN TURBULENT TIMES

How to thrive in uncertainty when it rains for a day or doesn't rain for a year.

For the first time in history, the world felt like a farmer. We the people of the land know about disruptions and delays to our way of life: equipment breaks, rain delays planting, storms taking out the harvest, or a seven-year drought hits. This time, the globe felt as one. With a few adjustments, many of the daily business practices and survival strategies we use as farmers and small business owners became key to helping the world become a better place.

I'm not going to debate the truth or falsehoods of the latest pandemic. I'm here to say that slow-downs, stoppages, and shifts happen for many reasons, sometimes for hours or years. We can weather the storm. The fact is business will go on and individuals who adapt to dynamic times will move forward despite uncertainty. Farmers plant seeds of faith and so should we. We have all created our new comfort zones including what we expect and accept.

It's time to pull back from the hype and see what is really happening around us. Personal life has always had influence over our business operations. Think about all the delays we encounter when operating for a season, much less for an entire year: known or unknown, planned or unplanned, short-term or long-term. Setbacks from lost keys and red lights to traffic jams are so prevalent that we don't notice them most of the time. They are the daily things we complain about. For bigger issues that shift life, the insurance industry came up with a title—these are QLE or qualifying life events. Some of us are facing harsh occasions now and continuing to endure. With a farmer's heart and an entrepreneurial spirit, we will move forward stronger than ever.

What you need to know about you and your customers' brains.

Farm and small business owners need to operate out of abundance, not fear. The economy has always expanded and contracted over time. Nowadays, the ebb and flow of this tide happens at an ever-increasing pace. As always, it's up to business owners to find the best ways to meet their customers' needs and deliver what people are willing to buy.

Know your ideal customer. Consumers are people, and owners need to be understanding to meet the patron where they are.

We engage our customers to build a community around our farm business. We need to be willing to adapt quickly and turn with the times. Mandates may come from all directions, and we are and will be required to follow those guidelines or shut down.

Befriending and partnering with other small businesses and similar farms help spread the risks as well as the rewards. You share in the benefits as you also minimize the risks. This is important when trying new markets and when keeping the learning curve low.

One of the best ways to understand what customers want is to ask them directly. During live events, have survey cards available, or if you're comfortable walking among the crowd, talk to people directly for their feedback.

Customer psychology has permanently changed.

People buy from people, and this is still true when it comes to online. Customers have power in the form of reviews. We don't know these people, but we're taking them at their word, for better or worse. These words become permanent in online profiles, and people know it. Individuals don't want to waste time or money, so they're more willing to start their research with reviews of previous patrons, good or bad. Doing your best as the owner to set expectations and meet those ideas for every customer can create a winning atmosphere.

Our in-person behavior has shifted, whether it's shopping, getting coffee, or hosting a child's birthday party. People see society differently now, and since humanity needs interaction, it's important to keep the community together while observing safety precautions.

Discover new tools to add to your trade.

E-commerce has become the lifeblood of business. Since its creation, the concept of e-business has flourished. Online retailing is critical for success when the in-person/brick and mortar economic wheels slow or stop for a time.

In the event you're not already taking advantage of e-commerce, considering current economic conditions, you might save the farm or at least add to your bottom line if you do so.

By putting your products online, your operation opens up to new markets and gains exposure to new sections of the public, all while taking better care of your current customers. Granted, if you don't have products, this is where finding a local farm that has value-added items like jams or pickles could be a winning situation for both farms.

Services can be added to your e-commerce in the form of classes. Teaching classes on-site can be digitized to present to a global market. At the same time you're developing your brand of education topics, you can decide to handle the details yourself or hire someone to do it. By taking what you know and giving back to your community, you are able to pivot during complex times. Once you've created a course, it's done and becomes passive income. Getting started can be simple; most cell phones offer an easy starting point. Although better equipment will yield enhanced results, that comes with more investment of time and money. I know plenty of people who are quite successful simply using their phones.

As you enter the online world of commerce, you'll find that successful individuals are data-driven. You can track just about everything online, which is helpful in calculating what is successful and what isn't. Patterns emerge when you start understanding the numbers. It can be as simple as comparing how many people indicate they're going to your event versus the actual attendance. By keeping numbers on similar items, you discover standards to run your business more efficiently. Having these statistics helps you plan for next year, find your best-selling products, or know when to buy in bulk to save money and lessen waste.

How do you get your ideal customers?

Marketing and advertising have always been keys for any business to survive tough times and thrive in sounder times. These leading business practices could mean life or death for your farm business.

Technology is the biggest factor leveling the landscape of today's business world. With creativity, you can expand your agritourism operation into the digital world. This shift will help you replace or expand your income during dips in the economy. A beautiful aspect of technology is the ability to 'farm' out most of the work to others who love doing those types of tasks. This fact allows you to do what you do best, build your operation.

To conduct the interviews for this book, with permission, I recorded them on my cell phone using the built-in app that came preinstalled. This allowed me to play back the conversation and annotate my notes at my own pace. Once you have content in this form, it's easy to transition to another type of media. Utilizing video to audio to print allows you to reach your audience where they consume content.

Whenever using recording devices for interviews, podcasting, or other content creation, become familiar with the hardware and software beforehand. The last thing you want to do is get in contact with a busy individual, start a conversation, and then have trouble recording it. Before I started the interviews, I tested out the app, talking with my wife who was in a room across the house.

Some technology questions can be handled by members of the younger generations residing in your house. Most apps or devices are going to have a customer service department that should be able to walk you through any snag. Strong customer service can be a legitimate selling point for any product over another. I tried a voice recording app that came with a transcription feature. I could send the track straight to the company, and for a fee, they would transcribe it.

Today, marketing and advertising are driven by social media in one form or another. Word of mouth is still the best method of becoming known; however, people don't interact as much as they used to. Whether our lives have gotten too busy or the latest health concerns have kept us apart, we still need to communicate. What was once face-to-face is now spoken over Wi-Fi. These digitally expressed words remain powerful at directing our dollars to a recommended business or away from a disappointing establishment. In my opinion, there's merit to the statement that your business can thrive or die on social media.

When you are known on any of the social platforms, it's critical to post regularly. With your particular operation, the intervals you post may vary from daily, weekly, or monthly, but consistency is the key. When you have a fan base, they'll be expecting reliable updates. This is not to say you can't post additional thoughts randomly, but to maximize your presence you will need consistent interactions.

In building an audience, you may notice a gathering of other farms and business owners. This is a good thing because two people can accomplish

more than one. By supporting these companies on their own platforms, you begin to form a relationship. Don't go along with people just to get something. If their values don't match yours, you won't want to associate with them because in time you could be viewed together.

Once you have gotten to know someone by joining their journey online and commenting on their feeds, you both could bring value to each other. This is when you start introducing your audience to the other operation. Here is where you need a solid abundance mindset. Although you might be in the same sector of agritourism, you need to know there is enough commerce to go around, as long as you stay true to your purpose and plan.

By following people and making worthwhile comments on their posts, you bring valuable content to their site. People notice not only the host but the followers as well. As you start becoming an active part of someone's community, their audience will begin to engage your digital profile.

Starting a newsletter with your warm audience is another way to stay relevant and in front of people. A weekly or monthly calling card can be your reason to let everyone know "What's Next on the Farm!" At the same time, you keep your people united with your operation and you give yourself a formal way to create content. Free software is available to make this task easier.

In public, display your farm logo — on everything from shirts, hats, even your vehicle. Put your brand on everything so people see you and your farm. This demonstrates how proud you are of what you're doing, and it's an easy conversation starter, then you can hand them a business card with a discount code to draw them out.

Let people know about your operation by talking about it. A lot of community organizations look for speakers to inform them of what's going on around their city or area. If you're not comfortable in front of a crowd, find self-help books and guided practices to become a better speaker. I taught school for a number of years and the stage never seemed to bother me; teach middle school and you can survive anything. I simply got up and spoke about what I knew, and I wound up bringing you this book. Always check with the event sponsor that it's okay to promote your farm and publicize upcoming events.

Meeting people in person at an event or conference is another way to broaden your audience. One on one, people can truly see your character and are more likely to become part of your digital following as an enthusiastic supporter.

The human race as a whole has been touched by a common factor and will forever be changed. Today is the best day to improve our situation. The landscape of business will continue to evolve at an increasing rate. These times are made for an abundant mindset, a farmer's heart, and an entrepreneur's spirit.

SECTION 4: CONCLUSIONS

Our farmers have a few last-minute words on agritourism.

Bonnie Chapa wants to remind readers to be flexible and recognize the difference between what you wish for and what's possible. Be realistic, yet dream to imagine where you're going. Always keep Mother Nature in mind; seasons and weather play major roles in anything agricultural. Build a strong support system of people who understand your vision; it will make everything a bit easier. Communicating your ambitions with this group is important for their understanding. Not everyone may want to be involved directly in your agritourism endeavor, but it's important they know what it is you're trying to accomplish.

Fran Tacy hopes you remember it's a business and must be treated as such, but no one ever said business couldn't be fun. Enjoy what you do to make a living.

Dewayne Hall says if you're interested in getting into or adding agritourism to your farm, go to two other venues and see how they're carrying on. Talk to the people running the business before you ever commit to joining them. Ask yourself:

What do they do?

What do you like about their operation?

What do you not like about their set-up?

Volunteering is a way to gain inside information on how a farm operates. Owners with a positive mindset will be grateful for the help and excited to help you in return. Individuals are the sum of their understanding, and we need everyone to educate new people with farm-related experiences; that's how we're going to make the world a better place.

Matt Wilkinson wrapped up his remarks with the idea that money is not the only form of payment. The look of gratitude on parents' faces when you show their child where real food comes from is definitely a form of payment and may be all you need. However, money will keep the lights on and the gate open so you can share your passion and love with others. Oh, the world we would have if we all paid it forward and shared our passion for things that grow.

John Moody finishes up by telling people to simply do it, no one can do it for you.

James Halvorson says you only get one opportunity to live your life, don't grow old regretting you never tried something. He says, "Agritourism is about fun people providing a fun experience to those who don't have that experience readily available." When we producers keep our nose to the grindstone, we often get lost in the mundane, but, when newcomers visit our operations, they see a whole new world and our farm becomes their personal reality.

James wants you to remember agritourism must be enjoyable for everyone, owner and customer alike. While it's necessary for the patron to have a good time, it's almost more important that you, the owner, find joy in providing possibilities. Visitors won't come back and owners won't continue if it's not fun for all.

Not all payments are monetary. He remembers the youngsters on his farm picking and pulling fresh vegetables for the first time, all with wide smiles as they ate carrots and broccoli right from the garden.

What I would like to leave you with is that you can be successful, you can do this, and, with agritourism, your farm dream can be even more than you know. We are all trying to educate about our passions, the land, farming, and more. We offer experiences to help patrons gain first-hand knowledge of skills required to stay on the farm. In this way, they'll know what it's like to be a bit more free. We want others to know what we love. It still amazes me how many agribusiness owners use the same words and phrases to describe what's going on behind the scenes and in their minds.

Whether you're adding to your current operation or starting from scratch, don't buy this book and think you're going to put on a pumpkin patch in two months when you've never worked one before. Be sure to put in the time to do your homework and volunteer. Nothing can replace hands-on involvement. It's better to spend a year doing research, start small, and grow the business correctly, rather than fail.

Part of your learning curve is to discover what it's like to move Christmas trees all day. You need to find out if that's the type of work you want to be doing, because while you can hire the labor, what happens when none of the muscle shows up? Agritourism may be demanding, but, by design, it can fit your business model. At the end of a day of volunteering, if you're tired but still excited and still love the work, then that type of agritourism may be the place for you.

I want *you* to get to the end of this book and say, "I can do this!"

With all the warnings about doing your research, the one thing to remember is you need to *do* something. A few will overanalyze until nothing gets done—they call that "analysis paralysis." So once your tasks are complete, take a step forward, and act. You only grow when you put seeds in the ground.

Agritourism can create a place for people to reconnect with the land, and for them to not only see where food comes from but for them to be empowered to grow food for themselves.

Thank you for being part of my agritourism story. I hope we connect soon. Until then …

P.S. Thank you to Bonnie, Fran, Dewayne, Matt, John, James, and all the people who contributed to bringing this book to life so it may live on as a part of your story.

PHOTO CREDITS

Front Cover
Clockwise from upper left: Adobe Stock (5)/Evrymmnt, jul14ka, Dmitriy Kapitonenko, EvgeniiAnd, alexkich

Introduction
Page 5: Adobe Stock (4)/byrdyak, 6–7: nareekam, 8: Pixel-Shot, 9: lara-sh

Profiles
Adobe Stock/10: maxbelchenko; 12–17: courtesy photos; 18: Charles Deviney, at the Robinson Family Farm, owners Brian and Helen Robinson, in Temple, TX; 19: Adobe Stock/Christin Lola

Chapter 1
Adobe Stock (2)/20–21: nataliaderiabina, 22: Tetatet; 27; Dewayne Hall; 28: Fran Tacy

Chapter 2
30: Dewayne Hall; Adobe Stock (2)/33: oksix. 34: Jillian; 36: Dewayne Hall

Chapter 3
Adobe Stock (2)/38: Matteo, 40: CK

Chapter 4
Adobe Stock (3)/42–43: yanadjan, 44: Vadym. 46: GAYSORN

Chapter 5
48: Adobe Stock (2)/48: musicphone1, 50: Leslie Rodriguez

Chapter 6
Adobe Stock (2)/52: maxdigi, 54: Halfpoint

Chapter 7
Adobe Stock (2)/56: Evelyne Mertens, 58: v_sot

Chapter 8
Adobe Stock/60: pressmaster

Chapter 9
62: Bonnie Chapa; Adobe Stock (2)/65: Sergii Mostovyi, 66: Maridav

Chapter 10
Adobe Stock (4)/68–69: Mary, 70: Mike, 72: BGStock72; 75: Zoltan Galantai

Chapter 11
Adobe Stock (2)/78: Robert Lerich, 81: mityru

Chapter 12
Adobe Stock (4)/84: Bonita. 87: Olga. 88: HollyHarry. 91: Yay Images

Chapter 13
94: Fran Tacy; 98: Bonnie Chapa; Adobe Stock/101: Fokke Baarssen

Chapter 14
Adobe Stock (2)/104: rolffimages, 107: Video_StockOrg

Conclusions
Adobe Stock/108–109: Trevor Parker Photo

Back Cover
Charles Deviney, at the Robinson Family Farm, owners Brian and Helen Robinson, in Temple, TX

NOTES